Fluency with Fractions

TEACHER'S GUIDE

Steph King

YEAR
5

Rising Stars UK Ltd
7 Hatchers Mews, Bermondsey Street, London, SE1 3GS

www.risingstars-uk.com

Published 2014
Reprinted 2014, 2015
Text, design and layout © Rising Stars UK Ltd.

Author: Steph King
Consultant: Cherri Moseley
Publisher: Fiona Lazenby
Project Manager: Debbie Allen
Editorial: Katharine Timberlake, Kate Manson
Cover design: Burville-Riley Partnership
Design: Marc Burville-Riley
Typesetting: Fakenham Prepress Solutions
Illustrations: Louise Forshaw / Advocate Art, Richard and Benjamin,
    Fakenham Prepress Solutions
CD-ROM development: Alex Morris

British Library Cataloguing in Publication Data.
A CIP record for this book is available from the British Library.

ISBN: 978-1-78339-184-4

Printed by: Ashford Colour Press Ltd, Gosport, Hants

Contents

Fractions in the 2014 National Curriculum

The National Curriculum aims to ensure that all pupils become fluent in the fundamentals of mathematics, can reason mathematically and can solve problems by applying their mathematics. With a significant shift in expectations in the 2014 Programme of Study, children are required to work with and calculate using a range of fractions at an earlier stage. Achieving fluency will depend on developing conceptual understanding through a variety of practical and contextual opportunities.

Statutory requirements and non-statutory guidance

At first glance, the statutory requirements for the Fractions domain for younger children may not appear to be that extensive. However, it is important to note that each 'objective' is made up of a range of different skills and knowledge that need to be addressed. We must remember that mastery of one aspect does not necessarily imply mastery of another.

The Programme of Study also provides non-statutory guidance that helps to clarify, secure and extend learning in each domain to best prepare children for the next stage of mathematical development. Units in this *Fluency with Fractions* series, therefore, also address some aspects of the non-statutory guidance. These objectives are flagged where applicable.

Fractions across the domains

Learning about fractions is not exclusive to the Fractions domain in the Programme of Study. Conceptual understanding of fractions is also addressed and applied through work on time, turns, angles and through many other aspects of measurement, geometry and also statistics. We must also remember to continue to practise and extend learning from previous year groups even if a concept is not explicitly covered in the Programme of Study for the current year group. The other domains provide useful opportunities for this.

Making the links: decimals, percentages, ratio and proportion

Children will first experience decimals in the context of measurement. However, security with place value is vital if they are to truly understand how the position of a digit on either side of the decimal point determines its size. Place value charts and grids are used in this series of books to continue to reinforce this concept and to help children make sense of tenths, hundredths and thousandths.

As children progress through the Programmes of Study, they will later meet percentages. Recognising that a fraction such as $\frac{25}{100}$ can be written as $25 \div 100$, and therefore as 0.25, will help make the connection to 25%.

Finding and identifying equivalent fractions will later pave the way for understanding equivalent ratios.

For this reason, within the *Fluency with Fractions* series, the Year 4 book includes work on decimals, Year 5 includes percentages and Year 6 goes on to incorporate ratio and proportion.

Developing conceptual understanding through the use of resources

Children should be given opportunities to develop conceptual understanding through a range of practical experiences and the use of visual representations to help them make sense of fractions. Manipulatives, such as Base 10 apparatus, cubes and counters, along with other resources, should be used skilfully to model concepts and provide a reference point to help children make connections for future learning. Moving in this way from concrete resources to pictorial representations to symbolic notation for fractions will help to secure conceptual understanding.

Developing mathematical language

Language is often cited as a barrier to learning, so it is important to model technical vocabulary that helps children to use it confidently and to help them explain their mathematical thinking and reasoning. Appropriate language structures are suggested throughout the Units.

Using representations to support understanding

Fractions is a part of mathematics that children often find more difficult to learn than other areas. This, in turn, is often the result of teachers finding the concepts more difficult to teach. We need to help children to see what we mean and make links to other familiar representations that they know, e.g. number lines.

Historically, images to support the teaching of fractions have tended to be related to real-life examples that children see 'cut-up' and shared. Pizzas, cakes and chocolate are examples of this. Although these representations are valuable (particularly circular images that will later inform work on pie charts), it is the linear image that directly relates to the number line that will support the transition of concrete to abstract when counting and calculating.

Throughout the *Fluency with Fractions* resources, fraction bar images are used in each year level to introduce the concept of fractions as equal parts of a whole, equivalence, counting (linked to the number line) and calculating.

The following diagrams provide a few generic examples to illustrate how different images are used. Templates for some useful images are provided on the accompanying CD-ROM.

- Fractions as equal parts of one whole.

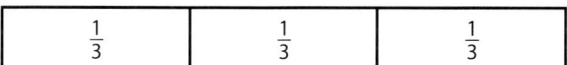

- Linking fractions to counting on a number line.

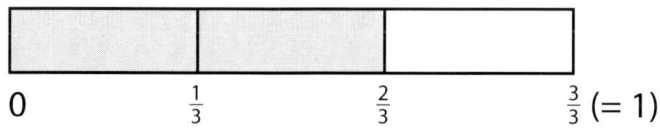

- Counting on the number line to reinforce that fractions are numbers in their own right. Counting paves the way for calculating.

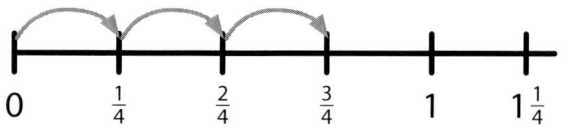

- Developing a range of images to explore equivalent fractions.

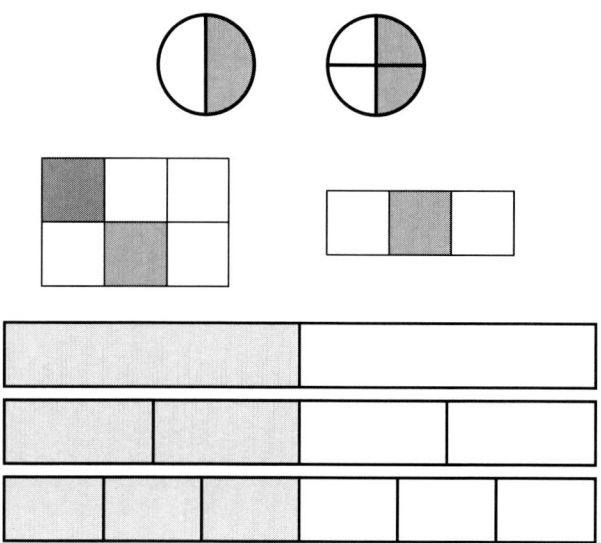

- Comparing fractions on a number line.

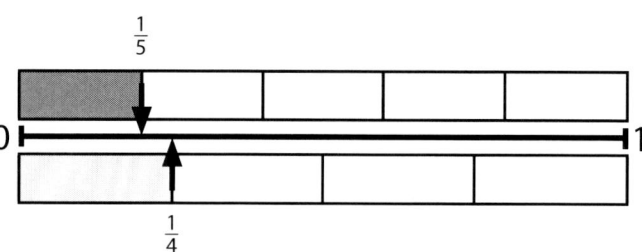

- Using fraction bars to support early calculation of fractions of amounts.

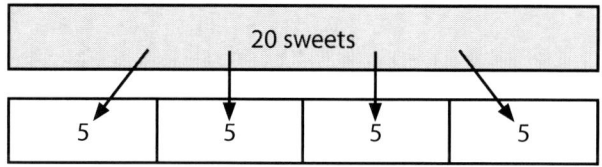

- Using fraction bars to support identifying an amount represented by a fraction.

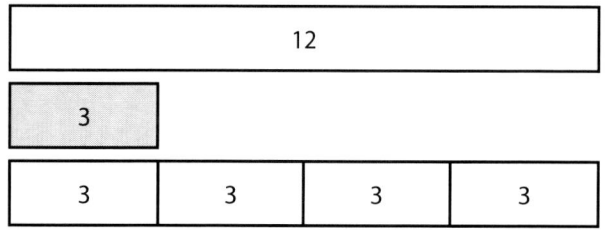

How to use this book

The Units in this book support the development of conceptual understanding of fractions and are intended to be used to introduce concepts. Learning should be practised and revisited regularly using other resources to consolidate and deepen understanding.

Each Unit within the books is structured in the same way, providing guidance to support teachers and an example teaching sequence.

Tasks can be used as suggested or adapted accordingly to meet the needs of each setting. Guided learning provides an opportunity for the adult to take learning forward with a group or to take part in an activity that has a greater problem-solving element and where language may be more demanding. Additional editable resource sheets are provided on the accompanying CD-ROM to support this.

Bold text shows the link to the NC objectives or the non-statutory guidance.

Please check that prior learning is in place before working on this unit.

This section helps teachers to make connections through the use of visual representations and language structures.

Tasks may be directed at the teacher to run the activity with children as guided learning; directed at the teacher to explain the activity to children to do independently; or directed at children to be photocopied and given out for independent work. Tasks increase in level of difficulty.

Task B is aimed at the majority of children who will progress at the expected rate.

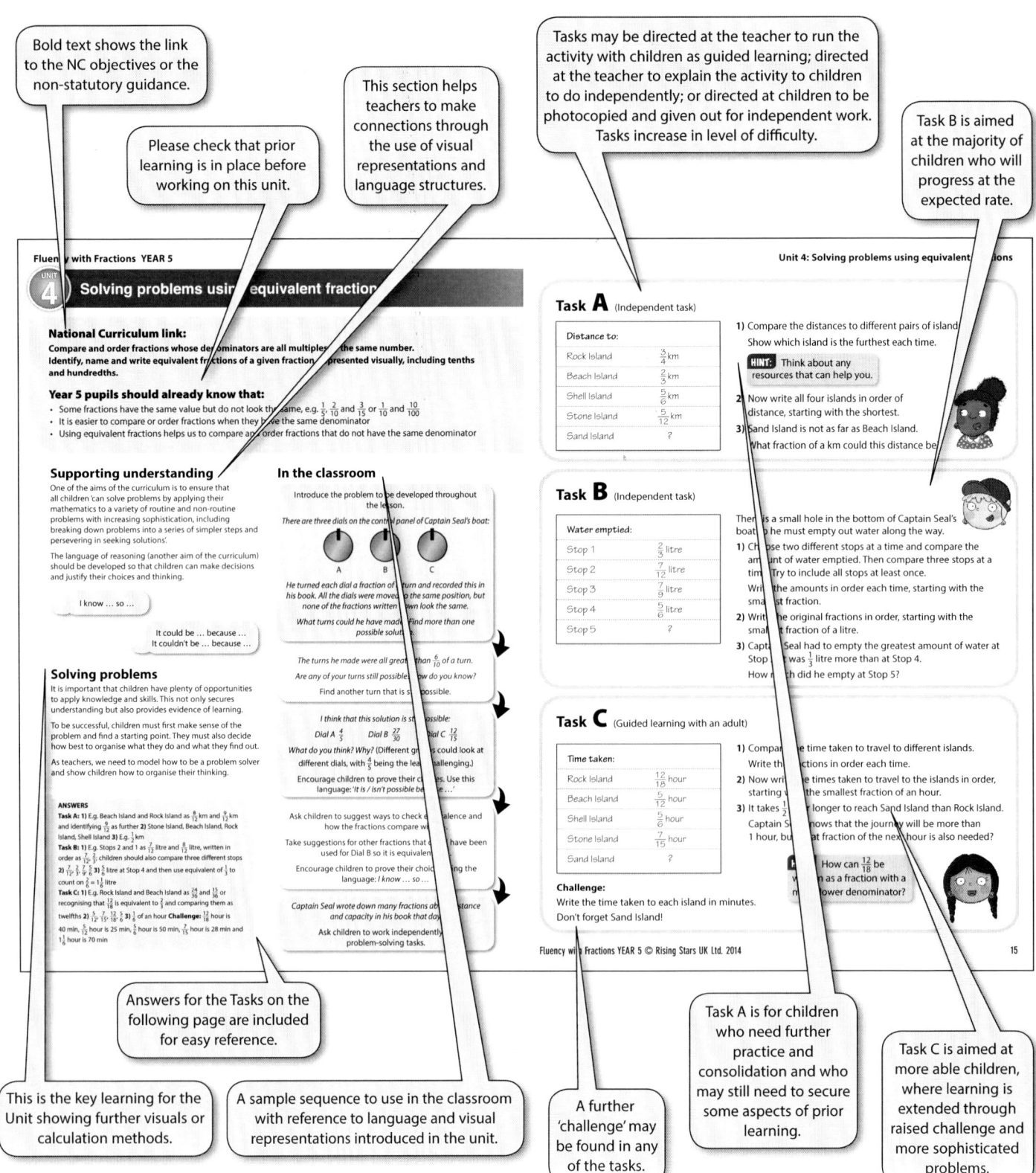

Answers for the Tasks on the following page are included for easy reference.

This is the key learning for the Unit showing further visuals or calculation methods.

A sample sequence to use in the classroom with reference to language and visual representations introduced in the unit.

A further 'challenge' may be found in any of the tasks.

Task A is for children who need further practice and consolidation and who may still need to secure some aspects of prior learning.

Task C is aimed at more able children, where learning is extended through raised challenge and more sophisticated problems.

Curriculum mapping grid

The grid below shows in which Units objectives from the 2014 National Curriculum Programme of Study for Year 5 are covered. Note that objectives are revisited regularly and learning progressed in subsequent units. In the National Curriculum link section of each Unit, bold text is used to indicate which specific part of the overarching objective is addressed within the Unit, since objectives often cover a range of different knowledge and skills (particularly for younger age groups).

Objectives	Unit 1	2	3	4	5	6	7	8	9	10	11	12	13	14	15	16	17	18	19	20	21	22
Identify, name and write equivalent fractions of a given fraction, represented visually, including tenths and hundredths.	✓		✓	✓																		
Continue to practise counting forwards and backwards in simple fractions.		✓																				
Extend counting from year 4, using decimals and fractions including bridging zero, for example on a number line.		✓									✓											
Compare and order fractions whose denominators are all multiples of the same number.				✓																		
Read and write decimal numbers as fractions [for example, $0.71 = \frac{71}{100}$].					✓																	
Recognise the per cent symbol (%) and understand that per cent relates to 'number of parts per hundred', and write percentages as a fraction with denominator 100, and as a decimal.						✓																
Recognise and use thousandths and relate them to tenths, hundredths and decimal equivalents.							✓															
Recognise mixed numbers and improper fractions and convert from one form to the other and write mathematical statements > 1 as a mixed number [for example, $\frac{2}{5} + \frac{4}{5} = \frac{6}{5} = 1\frac{1}{5}$].								✓														
Add and subtract fractions with the same denominator and denominators that are multiples of the same number.									✓	✓												
Add and subtract decimals, including a mix of whole numbers and decimals, decimals with different numbers of decimal places, and complements of 1 (for example, $0.83 + 0.17 = 1$).												✓										
Multiply proper fractions and mixed numbers by whole numbers, supported by materials and diagrams.													✓	✓								
Connect multiplication by a fraction to using fractions as operators (fractions of), and to division.													✓									
Interpret non-integer answers to division by expressing results in different ways according to the context, including with remainders, as fractions, as decimals or by rounding (for example, $98 \div 4 = \frac{98}{4} = 24\ r\ 2 = 24\frac{1}{2} = 24.5 \approx 25$).															✓							
Round decimals with two decimal places to the nearest whole number and to one decimal place.															✓							
Read, write, order and compare numbers with up to three decimal places.																✓						
Solve problems involving numbers with up to three decimal places.																	✓					
Recognise and describe linear number sequences, including those involving fractions and decimals, and find the term-to-term rule.																		✓				
Solve problems which require knowing percentage and decimal equivalents of $\frac{1}{2}, \frac{1}{4}, \frac{1}{5}, \frac{2}{5}, \frac{4}{5}$ and those fractions with a denominator of a multiple of 10 or 25.																			✓	✓		
Make connections between percentages, fractions and decimals (for example, 100% represents a whole quantity and 1% is $\frac{1}{100}$, 50% is $\frac{50}{100}$, 25% is $\frac{25}{100}$) and relate this to finding 'fractions of'.																					✓	✓

Equivalent fractions

National Curriculum link:
Identify, name and write equivalent fractions of a given fraction, represented visually, including tenths and hundredths.

Year 5 pupils should already know that:
- All fractions can be placed on the number line and some will sit in the same place as others
- Some fractions have the same value but do not look the same, e.g. $\frac{1}{5}$, $\frac{2}{10}$ and $\frac{3}{15}$ or $\frac{1}{10}$ and $\frac{10}{100}$

Supporting understanding
Language structures and images should be used to support children's understanding of equivalence.

> For every $\frac{1}{5}$ there are $\frac{2}{10}$ so $\frac{6}{10}$ is equivalent to $\frac{3}{5}$.

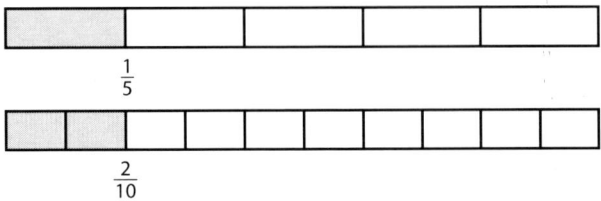

Children should be able to use similar language to explain other equivalences, e.g. for every $\frac{1}{10}$ there are 10 lots of $\frac{1}{100}$ or $\frac{10}{100}$. This means that for $\frac{3}{10}$ there are 3 lots of $\frac{10}{100}$, or $\frac{30}{100}$.

Equivalent fractions
Children will need to recognise a range of images that show equivalent fractions and answer questions about them:

Tim thinks that all of these images are equivalent to $\frac{3}{5}$. What do you think?

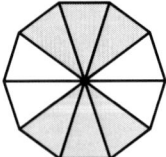 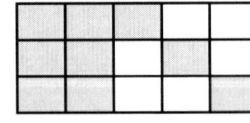 

What mistake has Tim made? Why do you think he did this?

How many more triangles should be shaded so the image is equivalent to $\frac{3}{5}$?

ANSWERS
Task A: 1) No **2)** No **3)** Equivalent to $\frac{1}{5}$ **4)** Equivalent to $\frac{1}{3}$ **5)** Shade two more sections in shape 1; shade one more section in shape 2
6) E.g. $\frac{4}{12}$, $\frac{5}{15}$
Task B: 1) Equivalent to $\frac{4}{5}$ **2)** Equivalent to $\frac{2}{3}$ **3)** No **4)** Equivalent to $1\frac{1}{5}$ **5)** Shade three more sections in shape 3 **6)** E.g. $\frac{4}{6}$, $\frac{8}{12}$, $\frac{10}{15}$
Task C: 1) E.g. 4 or 5 sixths; 4, 5, 6, 7 or 8 ninths; 4, 5, 6, 7, ... twelfths or fifteenths **2)** Any sets of equivalents that fit the criteria
3) Agree, because we need $\frac{4}{36}$ to be equivalent to $\frac{1}{9}$ as at least four sections must be shaded

In the classroom

Revisit the key point that all fractions can be placed on the number line and some will sit in the same place as others:

```
+----+--+----------------+--------+
0    1  1                3        1
     5  4                4
```

What other fractions sit in the same position as these on the number line?

Ask different groups to consider each fraction and find a way to prove their decisions. Use this language: *'For every $\frac{1}{5}$ there are $\frac{2}{10}$ so $\frac{6}{10}$ is equivalent to $\frac{3}{5}$.'*

Revisit using fraction bars to prove the equivalence of, say, $\frac{1}{5}$ and $\frac{2}{10}$, and then consider the relationship between the fractions, i.e. both the numerator and denominator have been multiplied by 2, therefore the relationship stays the same.

How do we know that $\frac{3}{15}$ and $\frac{4}{20}$ are also equivalent to $\frac{1}{5}$?

What has happened to the numerator and denominator this time?

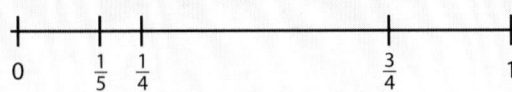

1	2	3	4	5	6	7
2	4	6	8	10	12	14
3	6	9	12	15	18	21
4	8	12	16	20	24	28
5	10	15	20	25	30	35

Confirm using the multiplication square, looking at the way the numbers are scaled each time.

Ask children to check the fractions they found in the earlier task and find others that are also equivalent.

Record as a string of equivalents, e.g. $\frac{3}{4} = \frac{6}{8} = \frac{9}{12} = \frac{12}{16}$, etc.

Introduce Tim's problem (left) or similar.

Ask children to discuss their ideas and check using the multiplication square, if needed.

Ask children who find this concept more difficult to focus on the decagon.

Task A (Independent task or guided learning with an adult)

Which of the shapes below match or are equivalent to the fractions on the number line?

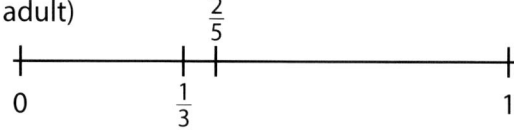

1) 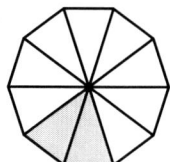   **2)**

3) 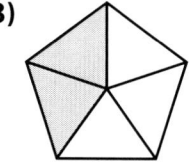   **4)**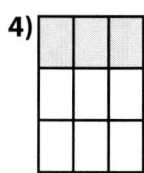

5) How can you change the incorrect shapes so they are equivalent to one of the fractions?

6) Now find some other fractions that are equivalent to $\frac{1}{3}$.

> **HINT:** Use multiplication tables to help you.

Task B (Independent task)

Which of the shapes below are equivalent to the fractions on the number line?

1)    **2)**   **3)**    **4)**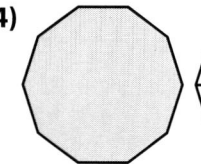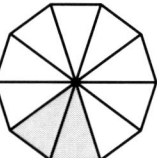

5) How can you change the incorrect shape so it is equivalent to one of the fractions?

6) Now find some other fractions that are equivalent to $\frac{2}{3}$.

> **HINT:** Use multiplication tables to help you.

Task C (Independent task)

Tim makes a set of shapes to show different equivalent fractions.

Each shape has a multiple of 3 sections.

Tim shades at least 4 sections on a shape each time.

1) Investigate to find fractions with different denominators that Tim can make.

2) List any equivalent fractions that can be made.

3) Do you agree with Tim's statement? Why?

> When there are less than **30** sections in my whole shape, I don't think I can find any fractions that are equivalent to $\frac{1}{9}$!

UNIT 2 Counting in fraction steps

National Curriculum link:

[Non-statutory guidance] **Continue to practise counting forwards and backwards in simple fractions.**
[Non-statutory guidance] **Extend counting from year 4, using decimals and fractions including bridging zero, for example on a number line.**

Year 5 pupils should already know that:

- All fractions can be placed on a number line
- We can count in fractions beyond 1 and use mixed numbers, e.g. $2\frac{1}{4}$, to help us
- Equivalent fractions and decimals have the same value and position on a number line, e.g. $\frac{2}{10}$, 0.2 and $\frac{1}{5}$

Supporting understanding

Children have been counting in fraction steps since Year 2, where it is featured in the non-statutory guidance.

Using images, such as fraction bars, can support children to understand counting in fractions beyond 1, e.g:

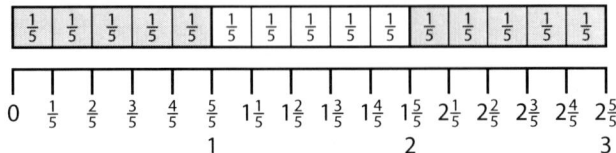

The number line can then be extended so the count can continue.

Questions such as: '*Using what you know about $\frac{2}{5}$, how can you explain where $10\frac{2}{5}$ or 10.4 and $11\frac{2}{5}$ or 11.4 are positioned on the number line?*' encourage children to use what they already know and make connections.

Counting in fractions on the number line

Counting in fractions also helps children make sense of mixed numbers and consider fraction and decimal equivalence.

In the example below, we want children to recognise that $\frac{2}{10}$, $\frac{4}{10}$, $\frac{6}{10}$, etc. have 'fifths' equivalents, and that $\frac{5}{10}$ is equivalent to $\frac{1}{2}$.

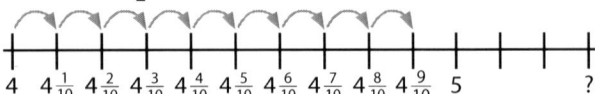

The number line can also be labelled in steps of 0.1.

Using the number line in this way will also support children to add and subtract fractions. *What do we know about the value of the question mark? How many tenths must I add to $4\frac{9}{10}$ to reach the question mark? Why can it not be 0.5?*

In the classroom

Practise counting in different fraction steps on a counting stick or using fraction bars. In unison, different groups should count in tenths from zero and the rest in decimal steps of 0.1. This will help to secure equivalent values.

Using knowledge of equivalence, children should discuss other fractions or decimals that could be included in this count, i.e. fifths, 0.2, halves and 0.5.

Now consider counting in tenths from any whole number, e.g. 10.

Pose the following, or similar, questions for children to consider:

- Using what you know about $\frac{3}{10}$, how can you explain where $10\frac{3}{10}$ will be positioned? How about $10\frac{2}{5}$ or 10.4?

- How many more tenths must I count to reach 11.4? How many fifths is this?

- How many fifths must I count back to reach 8.4? How many steps of 0.1 is this?

Introduce some examples of Tim, Katie and Ashton's counting. Ask children to explore and be prepared to explain their thinking, using number lines as proof.

Tim said: 'I counted four steps of $\frac{3}{10}$ on my number line. I started at 7. Where did I stop?

Katie said: 'I counted eight steps on my number line. I started at $6\frac{1}{2}$ and stopped on 7.3. What can you tell me about my step size?'

Ashton said: 'I started on $12\frac{5}{8}$ and counted back in steps of 0.25. Did I land on 10.5? How do you know?'

ANSWERS

Task A: Number lines showing that Tim landed on $\frac{1}{4}$ and Katie could have started on a whole number or a mixed number with $\frac{1}{2}$ as its fraction
Challenge: Steps of $\frac{1}{4}$
Task B: 1) Three different number lines starting from 3.4 and stopping on

e.g. 3.8 (two steps), 4.2 (four steps), 4.6 (six steps) **2)** Yes: 5.4 and 4.2
Task C: Tim: 10.8, 11.6, 12.4, 3.2, 14, 14.8, 15.6, 16.4, 17.2, 18, etc; Katie: yes, 14; Ashton: e.g. steps of $\frac{1}{2}$ from e.g. 13.5; steps of 0.1 from e.g. 10.7; steps of 0.2 from e.g. 10.6; steps of 0.6 from e.g. 10.4

Task A (Independent task)

Tim and Katie have been doing some more counting.

I started on $5\frac{3}{4}$ and counted back in halves. Prove that I did not land on zero.

I counted back in steps of **0.5**, but I did land on zero! How many ways can you make this true? What do you notice?

Draw number lines to show what you know about each child's counting.

Challenge: What step size could Tim have used so he did land on zero when he counted back from $5\frac{3}{4}$?

Task B (Independent task)

Ashton and Katie have been doing some more counting.

I counted forward in steps of $\frac{1}{5}$. I started on **3.4** and made an even number of steps. Where did I stop?

I started on **6** and counted back in steps of **0.3**. Can I land on the same number as Katie?

1) Draw three different number lines to show what you know about Katie's counting.

2) Explore Ashton's counting and find a way to answer his question.

Task C (Independent task)

Ashton, Katie and Tim have been doing some more counting.

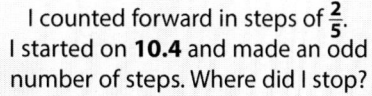

I counted forward in steps of $\frac{2}{5}$. I started on **10.4** and made an odd number of steps. Where did I stop?

Tim

I started on **19.25** and counted back in steps of $\frac{3}{4}$. Can I land on the same number as Tim?

Katie

I started on a different number and made different fraction steps. I had numbers from Katie and Tim's count in my count too! Can you work out what I did?

Ashton

Explore each child's counting to help find the answers to each of their questions.
Find a way to prove your thinking.

Comparing and ordering fractions

National Curriculum link
Compare and order fractions whose denominators are all multiples of the same number.

Year 5 pupils should already know that:
- Some fractions have the same value but do not look the same, e.g. $\frac{1}{5}$, $\frac{2}{10}$ and $\frac{3}{15}$ or $\frac{1}{10}$ and $\frac{10}{100}$
- It is easier to compare or order fractions when they have the same denominator
- A fraction can be rewritten as a division calculation, e.g. $\frac{1}{10}$ can be rewritten as $1 \div 10$ and $\frac{3}{4}$ can be rewritten as $3 \div 4$

Supporting understanding
Understanding equivalent fractions relies on children's security with multiplication and division facts and the concept of scaling, e.g:

3 made 2 times larger is written as 3×2.

12 made 2 times smaller is written as $12 \div 2$.

In the multiplication square shown here, we can clearly see that equivalents for the fraction $\frac{3}{4}$ can be found by scaling both the numerator and denominator by the same number.

The second column shows each number in the first column multiplied by 2, so $\frac{6}{8}$ is equivalent to $\frac{3}{4}$, while the third column shows each number in the first column multiplied by 3, so $\frac{9}{12}$ is also equivalent to $\frac{3}{4}$, and so on.

1	2	3	4	5	6	7	8	9	10	11	12
2	4	6	8	10	12	14	16	18	20	22	24
3	6	9	12	15	18	21	24	27	30	33	36
4	8	12	16	20	24	28	32	36	40	44	48
5	10	15	20	25	30	35	40	45	50	55	60
6	12	18	24	30	36	42	48	54	60	66	72
7	14	21	28	35	42	49	56	63	70	77	84
8	16	24	32	40	48	56	64	72	80	88	96
9	18	27	36	45	54	63	72	81	90	99	108
10	20	30	40	50	60	70	80	90	100	110	120
11	22	33	44	55	66	77	88	99	110	121	132
12	24	36	48	60	72	84	96	108	120	132	144

In Year 6, children will use the related knowledge of division and factors to help to simplify fractions.

Comparing and ordering fractions
We require children to reason about which fractions are easier to order and why. They should already know that fractions with the same denominator are easier to order, and may also recognise that those with the same numerator, e.g. $\frac{3}{8}$, $\frac{3}{6}$ and $\frac{3}{4}$, are easier to compare.

When neither numerators nor denominators are the same, we need to use knowledge of equivalent fractions, and decimals.

In Year 5, children compare and order fractions whose denominators are all multiples of the same number.

ANSWERS
Task A: 1) $\frac{1}{2} < \frac{7}{10}$ **2)** $\frac{2}{4} < \frac{5}{8}$ **3)** $\frac{7}{8} > \frac{3}{4}$ **4)** $\frac{5}{8}, \frac{3}{4}, \frac{7}{8}$
Task B: 1) $\frac{5}{9} < \frac{2}{3}$ **2)** $\frac{3}{4} > \frac{7}{12}$ **3)** $\frac{11}{12} > \frac{5}{6}$ **4)** $\frac{2}{3}$ is equivalent to $\frac{6}{9}$ so is smaller than $\frac{7}{9}$ **Challenge:** $\frac{7}{12}, \frac{2}{3} \left(\frac{8}{12}\right), \frac{3}{4} \left(\frac{9}{12}\right)$
Task C: E.g. $\frac{2}{3} > \frac{7}{12}$ using twelfths **Challenge:** Using what has already been found out or converting all fractions to the denominator 36; in order: $\frac{1}{6}, \frac{1}{3}, \frac{7}{18}, \frac{5}{9}, \frac{7}{12}, \frac{2}{3}, \frac{3}{4}, \frac{5}{6}$

In the classroom

Present three pairs of fractions for children to consider, e.g:

$\frac{7}{8}$ and $\frac{3}{8}$ $\frac{3}{6}$ and $\frac{3}{8}$ $\frac{4}{5}$ and $\frac{7}{10}$

Which fraction in each pair is the largest?

Was it easy to find the largest fraction in each pair? Which pair is more difficult? Why? (Fraction bars can be made available to support where necessary.)

Revisit the key point that fractions are easier to compare when they have the same denominator.

Consider the second pair of fractions and establish how these can also be compared easily because the numerator is the same.

Use models and images to support as necessary.

Return to $\frac{4}{5}$ and $\frac{7}{10}$.

Ask children to suggest their previous ideas and ways to solve the problem.

What do you notice about the denominator? How can this help us?

Establish that, as we know it is easier to compare fractions with the same denominator, we can use knowledge of equivalence to help us.

Use the multiplication square to model that $\frac{4}{5}$ can also be shown as $\frac{8}{10}$ because both the numerator and denominator have been multiplied by 2.

Now it is much easier to compare $\frac{7}{10}$ and $\frac{8}{10}$: $\frac{4}{5}$ is the largest fraction. Order the fractions.

Ask more able children to suggest another pair of fractions that can be compared in the same way.

This can also be confirmed using fraction bars or the key point that the fractions can be rewritten as $7 \div 10$ and $4 \div 5$.

Place value will support the decimal equivalent of $\frac{7}{10}$ as 0.7 and the calculator or knowledge that $\frac{1}{5}$ is 0.2 will help support $\frac{4}{5}$ as the decimal 0.8.

Task **A** (Independent task or guided learning with an adult)

Use a multiplication square and fraction bars to help you to compare these fractions.
Write your answers using the symbols < or > to help you.

1) $\dfrac{1}{2}$ $\dfrac{7}{10}$

2) $\dfrac{2}{4}$ $\dfrac{5}{8}$

3) $\dfrac{7}{8}$ $\dfrac{3}{4}$

4) Using what you know, order these fractions from smallest to largest.

$\dfrac{3}{4}$ $\dfrac{7}{8}$ $\dfrac{5}{8}$

Task **B** (Independent task or guided learning with an adult)

Use multiplication facts and fraction bars to help you to compare these fractions.
Write your answers using the symbols < or > to help you.

1) $\dfrac{5}{9}$ $\dfrac{2}{3}$

2) $\dfrac{3}{4}$ $\dfrac{7}{12}$

3) $\dfrac{11}{12}$ $\dfrac{5}{6}$

4) Using what you know, can you prove that $\dfrac{7}{9} > \dfrac{2}{3}$?

Challenge:

Can you order these three fractions? $\dfrac{7}{12}$ $\dfrac{3}{4}$ $\dfrac{2}{3}$

Check your decision using a calculator to convert each to a decimal.

Task **C** (Independent task)

1) Using multiplication facts, choose pairs of fractions with different denominators to compare.
Write your answers using the symbols < or > to help you.

$\dfrac{2}{3}$ $\dfrac{1}{6}$ $\dfrac{3}{4}$ $\dfrac{5}{6}$ $\dfrac{1}{3}$ $\dfrac{7}{12}$ $\dfrac{5}{9}$

Challenge:

How can you order all the fractions shown here?

What if this fraction is also included?

Check your decision using a calculator to convert each to a decimal.

$\dfrac{7}{18}$

UNIT 4 Solving problems using equivalent fractions

National Curriculum link:

Compare and order fractions whose denominators are all multiples of the same number.
Identify, name and write equivalent fractions of a given fraction, represented visually, including tenths and hundredths.

Year 5 pupils should already know that:

- Some fractions have the same value but do not look the same, e.g. $\frac{1}{5}$, $\frac{2}{10}$ and $\frac{3}{15}$ or $\frac{1}{10}$ and $\frac{10}{100}$
- It is easier to compare or order fractions when they have the same denominator
- Using equivalent fractions helps us to compare and order fractions that do not have the same denominator

Supporting understanding

One of the aims of the curriculum is to ensure that all children 'can solve problems by applying their mathematics to a variety of routine and non-routine problems with increasing sophistication, including breaking down problems into a series of simpler steps and persevering in seeking solutions'.

The language of reasoning (another aim of the curriculum) should be developed so that children can make decisions and justify their choices and thinking.

> I know … so …

> It could be … because …
> It couldn't be … because …

Solving problems

It is important that children have plenty of opportunities to apply knowledge and skills. This not only secures understanding but also provides evidence of learning.

To be successful, children must first make sense of the problem and find a starting point. They must also decide how best to organise what they do and what they find out.

As teachers, we need to model how to be a problem solver and show children how to organise their thinking.

ANSWERS

Task A: 1) E.g. Beach Island and Rock Island as $\frac{8}{12}$ km and $\frac{9}{12}$ km and identifying $\frac{9}{12}$ as further **2)** Stone Island, Beach Island, Rock Island, Shell Island **3)** E.g. $\frac{1}{2}$ km

Task B: 1) E.g. Stops 2 and 1 as $\frac{7}{12}$ litre and $\frac{8}{12}$ litre, written in order as $\frac{7}{12}$, $\frac{2}{3}$; children should also compare three different stops **2)** $\frac{7}{12}$, $\frac{2}{3}$, $\frac{7}{9}$, $\frac{5}{6}$ **3)** $\frac{5}{6}$ litre at Stop 4 and then use equivalent of $\frac{1}{3}$ to count on $\frac{2}{6}$ = $1\frac{1}{6}$ litre

Task C: 1) E.g. Rock Island and Beach Island as $\frac{24}{36}$ and $\frac{15}{36}$ or recognising that $\frac{12}{18}$ is equivalent to $\frac{2}{3}$ and comparing them as twelfths **2)** $\frac{5}{12}$, $\frac{7}{15}$, $\frac{12}{18}$, $\frac{5}{6}$ **3)** $\frac{1}{6}$ of an hour **Challenge:** $\frac{12}{18}$ hour is 40 min, $\frac{5}{12}$ hour is 25 min, $\frac{5}{6}$ hour is 50 min, $\frac{7}{15}$ hour is 28 min and $1\frac{1}{6}$ hour is 70 min

In the classroom

> Introduce the problem to be developed throughout the lesson.
>
> *There are three dials on the control panel of Captain Seal's boat:*
>
>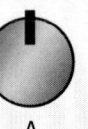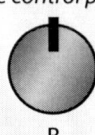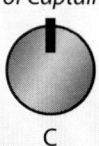
> A   B   C
>
> *He turned each dial a fraction of a turn and recorded this in his book. All the dials were moved to the same position, but none of the fractions written down look the same.*
>
> *What turns could he have made? Find more than one possible solution.*

> *The turns he made were all greater than $\frac{6}{10}$ of a turn.*
>
> *Are any of your turns still possible? How do you know?*
>
> Find another turn that is still possible.

> *I think that this solution is still possible:*
>
> *Dial A* $\frac{4}{5}$ *Dial B* $\frac{27}{30}$ *Dial C* $\frac{12}{15}$
>
> What do you think? Why? (Different groups could look at different dials, with $\frac{4}{5}$ being the least challenging.)
>
> Encourage children to prove their choices. Use this language: 'It is / isn't possible because …'

> Ask children to suggest ways to check equivalence and how the fractions compare with $\frac{6}{10}$.
>
> Take suggestions for other fractions that could have been used for Dial B so it is equivalent to $\frac{4}{5}$.
>
> Encourage children to prove their choices using the language: *I know … so …*

> *Captain Seal wrote down many fractions about distance and capacity in his book that day.*
>
> Ask children to work independently on problem-solving tasks.

Task A (Independent task)

Distance to:	
Rock Island	$\frac{3}{4}$ km
Beach Island	$\frac{2}{3}$ km
Shell Island	$\frac{5}{6}$ km
Stone Island	$\frac{5}{12}$ km
Sand Island	?

1) Compare the distances to different pairs of islands. Show which island is the furthest each time.

> **HINT:** Think about any resources that can help you.

2) Now write all four islands in order of distance, starting with the shortest.

3) Sand Island is not as far as Beach Island. What fraction of a km could this distance be?

Task B (Independent task)

Water emptied:	
Stop 1	$\frac{2}{3}$ litre
Stop 2	$\frac{7}{12}$ litre
Stop 3	$\frac{7}{9}$ litre
Stop 4	$\frac{5}{6}$ litre
Stop 5	?

There is a small hole in the bottom of Captain Seal's boat so he must empty out water along the way.

1) Choose two different stops at a time and compare the amount of water emptied. Then compare three stops at a time. Try to include all stops at least once.

Write the amounts in order each time, starting with the smallest fraction.

2) Write the original fractions in order, starting with the smallest fraction of a litre.

3) Captain Seal had to empty the greatest amount of water at Stop 5. It was $\frac{1}{3}$ litre more than at Stop 4.

How much did he empty at Stop 5?

Task C (Guided learning with an adult)

Time taken:	
Rock Island	$\frac{12}{18}$ hour
Beach Island	$\frac{5}{12}$ hour
Shell Island	$\frac{5}{6}$ hour
Stone Island	$\frac{7}{15}$ hour
Sand Island	?

1) Compare the time taken to travel to different islands. Write the fractions in order each time.

2) Now write the times taken to travel to the islands in order, starting with the smallest fraction of an hour.

3) It takes $\frac{1}{2}$ hour longer to reach Sand Island than Rock Island. Captain Seal knows that the journey will be more than 1 hour, but what fraction of the next hour is also needed?

> **HINT:** How can $\frac{12}{18}$ be written as a fraction with a much lower denominator?

Challenge:
Write the time taken to each island in minutes. Don't forget Sand Island!

Decimal numbers as fractions

National Curriculum link:

Read and write decimal numbers as fractions [for example, $0.71 = \frac{71}{100}$].

Year 5 pupils should already know that:

- Hundredths arise when dividing an object by 100 and dividing tenths by 10
- A fraction can be rewritten as a division calculation, e.g. $\frac{1}{10}$ can be rewritten as $1 \div 10$ and $\frac{1}{100}$ can be rewritten as $1 \div 100$
- Knowledge of tenths and hundredths helps us to work with decimals

Supporting understanding

The blank 100 square or base 10 equipment are powerful images to support understanding of decimals.

Each of the shaded strips is worth $\frac{1}{10}$ of the whole as it takes ten of these pieces to make up the whole.

Similarly, each of the shaded squares is worth $\frac{1}{100}$ of the whole as it takes a hundred of these pieces to make up the whole.

This visual also makes it easy to see that $\frac{1}{10}$ is equivalent to $\frac{10}{100}$.

This can also be confirmed using a place value grid.

Units or ones •	tenths	hundredths	
1			
0	1		$1 \div 10 = 0.1$
0	0	1	$1 \div 100 = 0.01$

The zero is used as a place holder so that the values of the other digits around it are known.

Decimal numbers as fractions

Using what we have found out from the place value grid, we can also count in decimal steps on the blank 100 square image. This confirms that $\frac{1}{10} = 0.1$ and $\frac{1}{100} = 0.01$.

Some children may already recognise $\frac{1}{100}$ as 1%.

This means that we can rewrite decimals as fractions.

The image here shows 6 tenths (0.6) and 4 hundredths (0.04).

We can also clearly see this as $\frac{64}{100}$ or 0.64, so $0.64 = \frac{64}{100}$.

In the classroom

Remind children that decimals and fractions are different ways of expressing numbers or proportions.

Use the blank 100 square or base 10 equipment to quickly re-establish that each of the shaded strips (or 10 sticks) is worth $\frac{1}{10}$ of the whole.

Ask children to use what they know about fraction and decimal equivalence to help to label each tenth as $\frac{1}{10}$, $\frac{2}{10}$, $\frac{3}{10}$, etc. and 0.1, 0.2, 0.3, etc.

How can we describe the value of each square as a fraction and a decimal?

Encourage children to explain how they can prove that $\frac{10}{100}$ is equal to $\frac{1}{10}$.

Using the image of $\frac{64}{100}$ (left), pose these, or similar, questions for different groups to consider:

- *How many tenths and hundredths can you see this time?*
- *How many more hundredths are needed to complete the whole? What facts did you use to help you?*
- *How can this fraction be written as a decimal? What calculation can be used to prove it?* (64 ÷ 100)

Confirm that the image shows $\frac{64}{100}$. Using a place value grid and the calculation $64 \div 100$, move the digits to prove the decimal equivalent.

So $0.64 = \frac{64}{100}$.

ANSWERS

Task B: $0.49 = \frac{49}{100}$, $\frac{5}{10}$ or $\frac{50}{100} = 0.5$, $0.52 = \frac{52}{100}$, $\frac{65}{100} = 0.65$, $0.66 = \frac{66}{100}$, $0.7 = \frac{7}{10}$ or $\frac{70}{100}$ **Challenge:** $0.82 = \frac{82}{100}$

Task C:

$\frac{27}{100}$	0.45	0.18
0.39	0.28	$\frac{23}{100}$
$\frac{24}{100}$	$\frac{17}{100}$	0.49

Top and bottom rows or left and right columns can be swapped

Challenge: No, because this will only affect the centre column and centre row; all other rows and columns will still sum to 0.9

Task A (Guided learning with an adult)

- Investigate the 100 square or base 10 equipment to ensure that children have a clear understanding of the way tenths and hundredths relate to the whole.
- Show images (e.g. those here) for children to re-create using base 10 equipment, counting in tenths and hundredths (including decimals) and recording the fraction and decimal shown.

  Ask questions such as:

  How many tenths have we counted?
  How many hundredths is this?

- Show each fraction on a place value grid with a zero indicating that there are no units.

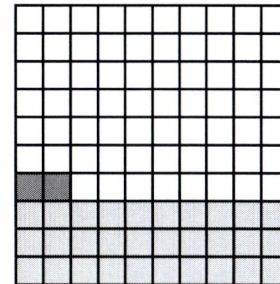 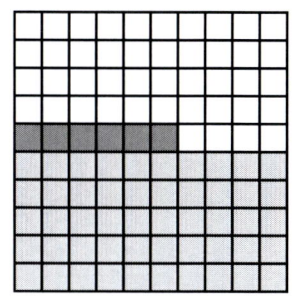

Units or ones •	tenths	hundredths
0	3	2

Task B (Independent task)

Tens	Units or ones •	tenths	hundredths

Use a place value grid to help to put these fractions and decimals in order from smallest to largest.

Remember to explain your decisions and write the decimal and fraction equivalent each time, e.g. $\frac{5}{10}$ = ___ .

?

Challenge:
Write the number that is $\frac{12}{100}$ larger than 0.7.
Remember to write the equivalent fraction.

$\frac{5}{10}$ 0.7   0.52

0.49   0.66 $\frac{65}{100}$

Task C (Independent task)

Fill the grid with these nine numbers so each row and each column sums to 0.9:

$\frac{17}{100}$ $\frac{23}{100}$ $\frac{24}{100}$ $\frac{27}{100}$

0.18   0.28   0.39   0.45   0.49

Challenge:
Pete thinks he can change the puzzle so each row and each column sums to 0.8 by subtracting 0.1 from the centre number. What do you think?

UNIT 6 Understanding and writing percentages in different ways

National Curriculum link:

Recognise the per cent symbol (%) and understand that per cent relates to 'number of parts per hundred', and write percentages as a fraction with denominator 100, and as a decimal.

Year 5 pupils should already know that:

- Decimals and fractions are different ways of expressing numbers or proportions
- Decimal numbers can be read and written as fractions, e.g. $0.71 = \frac{71}{100}$

Supporting understanding

91%	92%	93%	94%	95%	96%	97%	98%	99%	100%	$\frac{100}{100}$ or 1
81%	82%	83%	84%	85%	86%	87%	88%	89%	90%	$\frac{90}{100}$
71%	72%	73%	74%	75%	76%	77%	78%	79%	80%	$\frac{80}{100}$
61%	62%	63%	64%	65%	66%	67%	68%	69%	70%	$\frac{70}{100}$
51%	52%	53%	54%	55%	56%	57%	58%	59%	60%	$\frac{60}{100}$
41%	42%	43%	44%	45%	46%	47%	48%	49%	50%	$\frac{50}{100}$
31%	32%	33%	34%	35%	36%	37%	38%	39%	40%	$\frac{40}{100}$
21%	22%	23%	24%	25%	26%	27%	28%	29%	30%	$\frac{30}{100}$
11%	12%	13%	14%	15%	16%	17%	18%	19%	20%	$\frac{20}{100}$
1%	2%	3%	4%	5%	6%	7%	8%	9%	10%	$\frac{10}{100}$

Children should already have used the image of the blank 100 square to investigate place value for decimals and their fractional equivalents.

Per cent is described as 'number of parts per hundred'. Again, using the 100 square, we can begin to explore the relationship between $\frac{1}{100}, \frac{2}{100}, \frac{3}{100}$ and 1%, 2%, 3%, etc.

In the same way that we have previously described each small square as a hundredth because it takes 100 of them to complete the whole $\left(\frac{100}{100}\right)$, each is now also 1% as it takes 100 of them to complete the whole 100%.

> I know that … is equal to …% because …

Writing percentages in different ways

This image clearly relates half, with the equivalent fraction $\frac{50}{100}$, to percentages, i.e. $\frac{50}{100}\left(\text{or }\frac{1}{2}\right)$ to 0.5 or 50%.

Therefore, one more hundredth will equal $\frac{51}{100}$ or 0.51 or 51%.

Securing knowledge of decimal and percentage equivalents will later support calculations and help children to use calculators to work with percentages even when there is no % key.

Reminding children that $\frac{51}{100}$ can be written as the division $51 \div 100$ will support here.

In the classroom

Revisit the key point that decimals and fractions are different ways of expressing numbers or proportions.

Use the image of the blank 100 square to confirm that each small square represents $\frac{1}{100}$ of the whole or 0.01. Ask children to discuss and feed back anything they know about percentages.

Establish that percentages are another way of expressing proportions.

Describe the term *per cent* as 'number of parts per hundred'. Therefore each small square in the 100 square is 1%.

Quickly count together in 1% steps, alternating between decimals and fractions, as you go.

Shade $\frac{50}{100}$ of the 100 square.

Establish the equivalence of $\frac{1}{2}$ and 50%. Use this language: '*I know that $\frac{50}{100}$ is equal to 50% because it is made up of 50 small squares in the whole 100.*'

Invite children to use the image to come up with other fraction $\left(\frac{}{100}\right)$ and percentage equivalents of their own.

Encourage children to use the language: '*I know that … is equal to …% because …*'

Remind children that in the same way that we can write a decimal, such as 0.2 or 0.23 as $\frac{20}{100}$ or $\frac{23}{100}$, we can also write all our percentages and fractions with denominator 100 as decimals.

Use a place value grid to confirm that $\frac{23}{100}$ can be shown as a decimal using the division $23 \div 100$ and moving each digit two places to the right.

ANSWERS

Task A: 1) $\frac{25}{100}$, 25% **2)** $\frac{75}{100}$, 75% **3)** $\frac{60}{100}$, 60% **4)** $25 \div 100 = 0.25$, $75 \div 100 = 0.75$, $60 \div 100 = 0.6$

Task B: 1) $\frac{75}{100}$, 75% **2)** $\frac{80}{100}$, 80% **3)** $\frac{33}{100}$, 33% **4)** $\frac{95}{100}$, 95%
5) $75 \div 100 = 0.75$, $80 \div 100 = 0.8$, $33 \div 100 = 0.33$, $95 \div 100 = 0.95$
6) $\frac{45}{100}$, 0.5, 55%, $\frac{60}{100}$ or $\frac{6}{10}$

Task C: 1) $\frac{69}{100}$, 69%, 0.69 **2)** $\frac{23}{100}$, 23%, 0.23 **3)** 31%, $\frac{31}{100}$, 0.31; 77%, $\frac{77}{100}$, 0.77 **4)** 37%, 0.21, $\frac{5}{100}$ **5)** $\frac{8}{100}$, 0.36, 43%, $\frac{50}{100}$
6) Knowledge that $50 \times 2 = 100$, so he doubled each part of the fraction to find an equivalent with denominator 100

Task **A** (Independent task)

What does each image represent?

Write the fraction $\left(\frac{?}{100}\right)$ and percentage equivalents.

1) 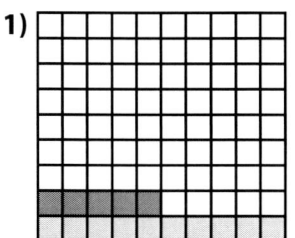 2) 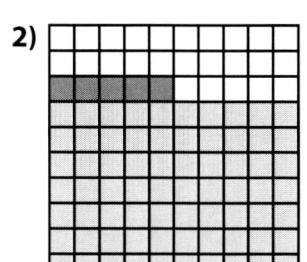 3)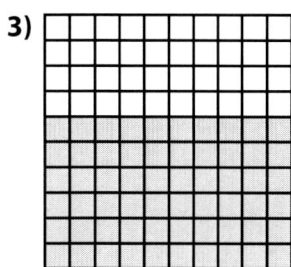

Remember that a fraction, e.g. $\frac{50}{100}$, can be written as $50 \div 100$, to help us to find the decimal equivalent as 0.5.

4) Write each of the fractions using division and then find the decimal equivalent.

HINT: A place value chart will also help you to find the decimal equivalent.

Task **B** (Independent task)

What does each image represent?

Write the fraction $\left(\frac{?}{100}\right)$ and percentage equivalents.

1) 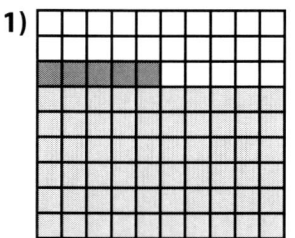 2)  3) 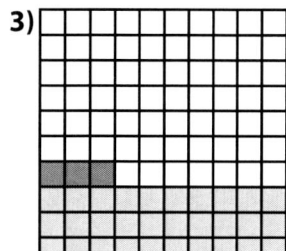 4)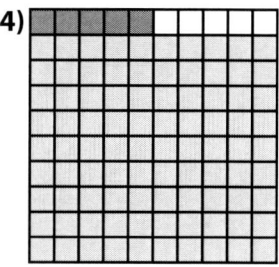

Remember that a fraction, e.g. $\frac{55}{100}$, can be written as $55 \div 100$, to help us to find the decimal equivalent as 0.55.

5) Write each of the fractions using division and then find the decimal equivalent.

6) Copy and complete this sequence: 25%, $\frac{30}{100}$, 0.35, 40%, ___, ___, ___, ___ .

Task **C** (Independent task)

1) 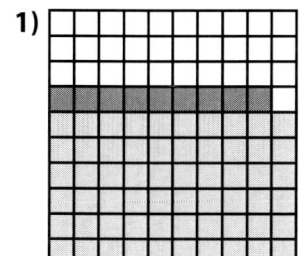 2)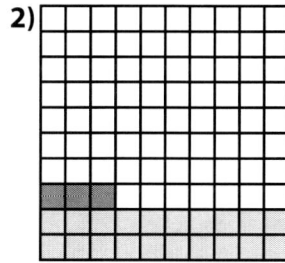

What does each image represent?

Write the fraction $\left(\frac{?}{100}\right)$, percentage and decimal equivalents.

3) What percentage of each shape is not shaded? Write the fraction and decimal equivalents.

Copy and complete these sequences:

4) 85%, 0.69, $\frac{53}{100}$ ___, ___, ___

5) ___, 0.15, 22%, $\frac{29}{100}$ ___, ___, ___

6) Tim wrote $\frac{45}{50} = 90\%$. What did he use to help him?

Recognising and using thousandths

National Curriculum link:

Recognise and use thousandths and relate them to tenths, hundredths and decimal equivalents.

Year 5 pupils should already know that:

- Hundredths arise when dividing an object by 100 and dividing tenths by 10
- A fraction such as $\frac{1}{10}$ can be rewritten as $1 \div 10$ and $\frac{1}{100}$ can be rewritten as $1 \div 100$
- Knowledge of tenths and hundredths help us to work with decimals

Supporting understanding

In Year 4, children used the context of money to help make sense of tenths and hundredths and their decimal equivalents.

A pound is divided by 10, resulting in 10 equal parts. Each tenth is worth 10 pence.

Each tenth (10p) is then divided by 10 again, resulting in 100 equal parts.

Each hundredth is worth a pence (1p) or £0.01.

This can also be confirmed using a place value grid.

Units or ones •	tenths	hundredths	
1			
0	1		$1 \div 10$
0	0	1	$1 \div 100$

The zero is used as a place holder so that the values of the other digits around it are known.

Recognising and using thousandths

Measurement can also support understanding of thousandths, e.g. 1 m is $\frac{1}{1000}$ of 1 km because it takes 1000 metres to equal 1 kilometre.

Units or ones •	tenths	hundredths	thousandths	
1				
0	0	0	1	$1 \div 1000$

So $1 \div 1000$ is equal to $\frac{1}{1000}$ or 0.001.

In the classroom

Revisit the key point that hundredths arise when dividing an object by 100 and dividing tenths by 10.

How does this information help us to describe thousandths?

Ask children to discuss and provide explanations.

Establish that thousandths arise when dividing an object by 1000 and dividing hundredths by 10 (or tenths by 100).

Use the place value grid to model dividing 1 by 1000 and that the same result can be made by dividing $\frac{1}{100}$ by 10, etc.

Reinforce how zero is used as a place holder so that the values of the other digits around it are known.

Record findings as e.g. $1 \div 1000 = \frac{1}{1000} = 0.001$.

Contextualise the concept by linking to measurement.

Use 1 km or 1 m as the whole so that $\frac{1}{100}$ is 10 m or 1 cm and $\frac{1}{1000}$ is 1 m or 1 mm, respectively.

Pose these, or similar, questions for different groups to consider (these are about kilometres):

- *How many lots of 1 m are in 1 km?*
- *What fraction of a km is 1 m? How do you know?*
- *What fraction of a km is 10 m? How many thousandths is this?* $\left(\frac{10}{1000}\right)$

Ask different groups to find $\frac{1}{1000}$ of each of these measurements using the place value grid to support:

2 km   5 km   12 km

However, secure tenths and hundredths for children needing additional consolidation.

Task **A** (Independent task or guided learning with an adult)

Units or ones •	tenths	hundredths	thousandths

Take a digit card and put it on the place value grid in the 'ones' position.

Divide the number by 100. Remember that you will need to move the digit card!

Record in a number sentence using the ÷ symbol.

Now find out what happens when the number is divided by 1000.

Repeat with different digit cards.

Challenge:

Sami wrote $7 \div 1000 = \frac{7}{100}$.

How do you know that he has made a mistake?

Use the language: 'Sami has made a mistake because ___ .'

Task **B** (Independent task)

Copy and compete the following:

1) $\frac{1}{1000}$ of 3 km = ___

2) $\frac{1}{1000}$ of ___ = 6 m

3) $\frac{10}{1000}$ of 1 km = ___

4) $9 \div 1000$ = ___

5) 0.007 = ___

6) There are ___ thousandths in $\frac{3}{100}$

In the long jump event, Pete jumps $\frac{2}{1000}$ km further than Tim. Tim jumps 5 m.

7) How long was Pete's jump as a fraction of a km?

8) Write Pete's jump as a decimal. Remember to write the unit of measurement.

Task **C** (Independent task)

HINT: Check the unit of measurement each time.

Copy and compete the following:

1) $\frac{3}{1000}$ of 1 km = ___

2) $\frac{8}{1000}$ of ___ = 8 mm

3) $\frac{15}{1000}$ of 1 kg = ___

4) $14 \div 1000$ = ___

5) 0.015 = ___

6) There are $\frac{60}{1000}$ in ___ hundredths

Abi throws the javelin $\frac{45}{1000}$ km in a competition.

Her second throw is 11 m longer and her third throw is a massive 0.063 km.

7) How far was Abi's second throw as a decimal?

8) What is the difference between her second and third throws? Give your answer as $\frac{?}{1000}$ km.

UNIT 8
Mixed numbers and improper fractions

National Curriculum link:

Recognise mixed numbers and improper fractions and convert from one form to the other and write mathematical statements > 1 as a mixed number [for example, $\frac{2}{5} + \frac{4}{5} = \frac{6}{5} = 1\frac{1}{5}$].

Year 5 pupils should already know that:

- We can count forward and backward in fraction steps beyond 1
- A whole can also be described as a fraction, e.g. fifths means there are five equal parts so the whole can be described as $\frac{5}{5}$

Supporting understanding

In Key Stage 1, children begin to combine fractions to make wholes. They later use the notation $\frac{2}{2}, \frac{3}{3}, \frac{4}{4}, \frac{5}{5}$, etc. to show that these fractions are all equivalent to one whole.

This underpins children's understanding of improper fractions and how they can be converted to mixed numbers.

Consider the counting that children will already have been doing, e.g. $\frac{1}{5}, \frac{2}{5}, \frac{3}{5}, \frac{4}{5}, \frac{5}{5}$, etc, and then continue in improper fractions:

| $\frac{1}{5}$ | $\frac{2}{5}$ | $\frac{3}{5}$ | $\frac{4}{5}$ | $\frac{5}{5}$ | $\frac{6}{5}$ | $\frac{7}{5}$ | $\frac{8}{5}$ | | |

We can then use knowledge of $\frac{5}{5}$ and a whole equivalence to place the 'whole' over the first $\frac{5}{5}$ counted and then start again from $\frac{1}{5}$:

| 1 | | | | | $\frac{1}{5}$ | $\frac{2}{5}$ | $\frac{3}{5}$ | | |

Converting between improper fractions and mixed numbers

Once the concept is secure, children can use knowledge of partitioning to help to convert improper fractions to mixed numbers and vice versa, e.g:

$\frac{13}{10} = \frac{10}{10} + \frac{3}{10} = 1\frac{3}{10}$

$\frac{17}{5} = \frac{5}{5} + \frac{5}{5} + \frac{5}{5} + \frac{2}{5} = 3\frac{2}{5}$, or $\frac{17}{5} = \frac{15}{5} + \frac{2}{5} = 3\frac{2}{5}$ for children who require fewer steps

$2\frac{4}{5} = \frac{5}{5} + \frac{5}{5} + \frac{4}{5} = \frac{14}{5}$, or $2\frac{4}{5} = \frac{10}{5} + \frac{4}{5} = \frac{14}{5}$

ANSWERS

Task A: 1) $\frac{6}{5} = 1\frac{1}{5}$ **2)** $\frac{9}{5} = 1\frac{4}{5}$ **3)** $\frac{5}{3} = 1\frac{2}{3}$ **4)** $\frac{12}{10} = 1\frac{2}{10}$ (or $1\frac{1}{5}$)

Challenge: It is also $\frac{12}{10}$; each step is double the size so only half as many are needed to be taken to reach the same place, or $12 \times 1 = 12$ and $6 \times 2 = 12$

Task B: 1) $\frac{12}{10} = 1\frac{2}{10}$ or $1\frac{1}{5}$ **2)** $\frac{11}{4} = 2\frac{3}{4}$ **3)** $\frac{14}{6} = 2\frac{2}{6}$ or $2\frac{1}{3}$; Pete: no, it is the same as $3\frac{1}{5}$, while $2\frac{1}{5}$ is $\frac{11}{5}$; Sally: $\frac{11}{3}$; Ahmed: $\frac{9}{5}$ as this is $\frac{18}{10}$, which is $1\frac{8}{10}$

Task C: Children may simplify fractions: **1)** $\frac{25}{6}$, $4\frac{4}{6}$; rule $+\frac{3}{6}$ **2)** 4, $\frac{13}{4}$, $2\frac{2}{4}$; rule $-\frac{3}{4}$ **3)** $4\frac{3}{9}$, $\frac{43}{9}$, $5\frac{2}{9}$; rule $+\frac{4}{9}$ **Challenge:** 0, $-3\frac{1}{3}$, $-\frac{15}{3}$; rule $-\frac{5}{3}$

In the classroom

Revisit the key point that we can count in fractions beyond 1 by counting in e.g. fifths from 0 to $1\frac{3}{5}$.

Record the count and make links to any equivalent fractions as you go, e.g. to $\frac{5}{5}$ and a whole or tenths, if appropriate.

Introduce the fraction $\frac{8}{5}$. Ask what is different about this fraction. Establish that we call fractions of this nature 'improper'.

Suggest that the same count can be made in steps of $\frac{1}{5}$, but this time finishing at $\frac{8}{5}$.

Will this count finish in the same place as before? How can you explain your decision? (Support discussions, where necessary, to establish that the count will also go beyond $\frac{5}{5}$ or 1.)

Take feedback and confirm using fraction bars (see left). Establish that $\frac{8}{5}$ has been split (or *partitioned*) into 1 whole $\left(\frac{5}{5}\right)$ and $\frac{3}{5}$. Record as $\frac{8}{5} = \frac{5}{5} + \frac{3}{5} = 1\frac{3}{5}$.

Ask different groups to work with the following, or similar, counts to practise converting improper fractions to mixed numbers. Make fraction bars available.

Counting in quarters to $\frac{7}{4}$

Counting in sixths to $\frac{11}{6}$

Counting in sixths to $\frac{19}{6}$

Pose these, or similar, questions for different groups to consider:

- *I counted in quarters up to $1\frac{1}{4}$. How can I write this as an improper fraction?*

- *How can I write $1\frac{3}{8}$ as an improper fraction?*

- *Is $2\frac{4}{7}$ equivalent to $\frac{11}{7}$? What mistake have I made?*

Task A (Independent task)

You will need at least two copies of each of these fraction bars.

1) Count 6 steps of $\frac{1}{5}$.

Cover $\frac{5}{5}$ with the 'whole' bar to help you to write the fraction as a mixed number.

Now try each of these counts.

Record where you land as an improper fraction and a mixed number.

2) 9 steps of $\frac{1}{5}$.

3) 5 steps of $\frac{1}{3}$.

4) 12 steps of $\frac{1}{10}$.

Challenge:

Count 6 steps of $\frac{2}{10}$. What do you notice about this count and the one you did in question 4? Why do you think this happened?

Task B (Independent task)

Fraction bars should be made available.

Complete these counts. Record where you land each time as an improper fraction and then use partitioning to show it as a mixed number.

1) Count 6 steps of $\frac{2}{10}$ **2)** Count 11 steps of $\frac{1}{4}$ **3)** Count 7 steps of $\frac{2}{6}$

Pete, Sally and Ahmed have also been doing some counting. Answer their question each time.

Pete  I landed on $\frac{16}{5}$. That is the same as $2\frac{1}{5}$. Am I correct?

 **Sally** I landed on $3\frac{2}{3}$. How many thirds have I counted? How do I write this as an improper fraction?

Ahmed  Which is larger: $1\frac{7}{10}$ or $\frac{9}{5}$?

Task C (Independent task or guided learning with an adult)

Fraction bars should be made available.

Complete these sequences using alternate improper fractions and mixed numbers.

What is the rule each time?

1) $\frac{13}{6}$, $2\frac{4}{6}$, $\frac{19}{6}$, $3\frac{4}{6}$, ___ , ___ .

2) $\frac{25}{4}$, $5\frac{1}{2}$, $\frac{19}{4}$, ___ , ___ , ___ .

3) $\frac{27}{9}$, $3\frac{4}{9}$, $\frac{35}{9}$, ___ , ___ , ___ .

4) Make up a sequence of your own starting from $\frac{24}{7}$.

Challenge:

Complete the sequence and write the rule: $\frac{10}{3}$, $1\frac{2}{3}$, ___ , $-\frac{5}{3}$, ___ , ___ .

UNIT 9

Adding fractions

National Curriculum link:

Add and subtract **fractions with the same denominator and denominators that are multiples of the same number.**

Year 5 pupils should already know that:

- A whole can also be described as a fraction, e.g. $\frac{5}{5}$. This helps us to convert an improper fraction to a mixed number, e.g. $\frac{7}{5}$ to $1\frac{2}{5}$
- It is much easier to add fractions when the denominators are the same

Supporting understanding

In previous units, children have counted in different fraction steps. Images have been used to support this understanding.

Fraction bars can be used to support counting on. It also supports the transition to number lines, e.g:

count on $\frac{3}{7}$ from $\frac{6}{7}$ as $\frac{6}{7} + \frac{3}{7}$.

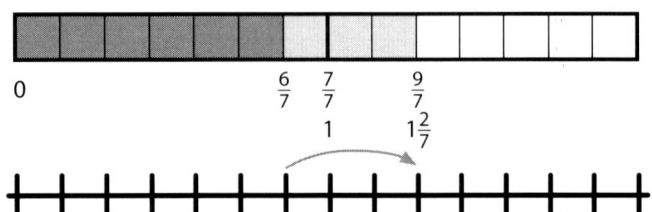

Knowledge of mixed numbers and improper fractions is vital here.

Recording addition calculations on the number line

The number line will also support the understanding of calculations that bridge the whole, e.g. $\frac{5}{8} + \frac{4}{8}$ by partitioning $\frac{4}{8}$ into $\frac{3}{8}$ (to complete the whole) and $\frac{1}{8}$.

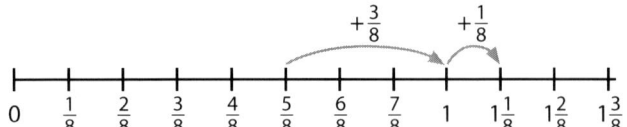

We can also use knowledge of improper fractions and the previous work on fraction bars to help us to complete the same calculation, e.g:

$\frac{5}{8} + \frac{4}{8} = \frac{9}{8} = \frac{8}{8} + \frac{1}{8} = 1\frac{1}{8}$

In the following example we must also use knowledge of equivalence, as we know that it is easier to add fractions with the same denominator:

$\frac{9}{10} + \frac{4}{5} = \frac{9}{10} + \frac{8}{10} = \frac{17}{10} = 1\frac{7}{10}$.

In the classroom

Refer back to the counting children have been doing and how they have converted from improper fractions to mixed numbers and vice versa.

Suggest a count in steps of $\frac{3}{8}$ starting from $\frac{7}{8}$.

Ask children to quickly find the first number we land on and any other numbers that will be in the count, expressing them as improper fractions and mixed numbers. Support groups where needed with reference to fraction bars.

Return to the first jump and record this on a number line, giving attention to bridging one whole.

Record the matching addition sentence as:

$\frac{7}{8} + \frac{3}{8} = \frac{10}{8} = \frac{8}{8} + \frac{2}{8} = 1\frac{2}{8}$.

How can we write the fraction remainder in a different way? $\left(\frac{1}{4}\right)$

Consider the second jump. Establish that we have to add another $\frac{3}{8}$ to $1\frac{1}{4}$. Record as $1\frac{1}{4} + \frac{3}{8}$.

What would it be sensible to do? Why?

Establish that using the original equivalent of $1\frac{2}{8}$ would be easier as then both denominators will be the same.

Confirm that $\frac{1}{4}$ and $\frac{2}{8}$ are equivalent using fraction bars.

Different groups should practise with the following calculations:

$\frac{9}{10} + \frac{4}{5}$

$1\frac{2}{3} + \frac{5}{6}$

$3\frac{4}{7} + 1\frac{3}{14}$

ANSWERS

Task A: 1) $1\frac{2}{5}$ **2)** $1\frac{1}{8}$ **3)** $1\frac{1}{3}$ **4)** $1\frac{1}{10}$ **5)** E.g. $\frac{3}{6} + \frac{4}{6} = 1\frac{1}{6}$ **6)** $1\frac{1}{4}$ **7)** $1\frac{2}{10} \left(1\frac{1}{5}\right)$ **8)** $\frac{7}{8}$

Task B: 1) $1\frac{1}{10}$ **2)** $2\frac{3}{8}$ **3)** $2\frac{3}{6} \left(2\frac{1}{2}\right)$ **4)** $3\frac{2}{10} \left(3\frac{1}{5}\right)$ **5)** E.g. $\frac{5}{6} + \frac{1}{12} = \frac{10}{12} + \frac{1}{12} = \frac{11}{12}$

6) E.g. $\frac{2}{3} + \frac{7}{9}$

Task C: 1) $2\frac{8}{9}$ **2)** $\frac{8}{18} + \frac{15}{18} = \frac{23}{18} = 1\frac{5}{18}$ **3)** $4\frac{5}{12}$ **4)** $6\frac{10}{12}$ or $6\frac{5}{6}$ **5)** E.g. $\frac{7}{8} + \frac{2}{3} + \frac{5}{6}$

$= \frac{57}{24} = 2\frac{9}{24}$ or $2\frac{3}{8}$, $\frac{3}{4} + \frac{2}{3} + \frac{5}{8} = \frac{49}{24} = 2\frac{1}{24}$; look out for examples where he should have used e.g. eighteenths (the question says that he had to use twenty-fourths) **6)** Children's own example, e.g. with thirds, fifths and sixths and needing to convert to thirtieths

Task A (Independent task)

You will need at least two copies of each of the fraction bars, up to tenths.
Find the answers to the following addition calculations:

1) $\frac{4}{5} + \frac{3}{5} =$ ____ **2)** $\frac{5}{8} + \frac{4}{8} =$ ____ **3)** $\frac{2}{3} + \frac{2}{3} =$ ____ **4)** $\frac{7}{10} + \frac{4}{10} =$ ____

5) Now make an addition calculation of your own using sixths.

Use the fraction bars to help you to find equivalent fractions for these additions:

6) $\frac{3}{4} + \frac{1}{2} =$ ____ **7)** $\frac{7}{10} + \frac{1}{2} =$ ____ **8)** $\frac{3}{4} + \frac{1}{8} =$ ____

Task B (Independent task)

You may need at least two copies of each of the fraction bars, up to twelfths.
Find the answers to the following addition calculations:

1) $\frac{4}{5} + \frac{3}{10} =$ ____ **2)** $1\frac{5}{8} + \frac{3}{4} =$ ____ **3)** $\frac{5}{3} + \frac{5}{6} =$ ____ **4)** $2\frac{1}{2} + \frac{7}{10} =$ ____

5) Now make an addition calculation of your own using sixths and twelfths.

6) Ami had to convert one of her fractions to ninths to help her with this calculation.
What calculation did she do?

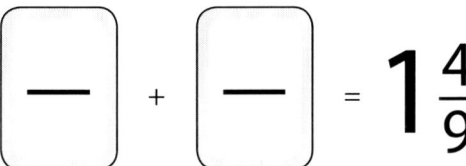

Task C (Independent task)

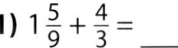

Find the answers to the following addition calculations:

1) $1\frac{5}{9} + \frac{4}{3} =$ ____ **2)** $\frac{4}{9} + \frac{5}{6} =$ ____ **3)** $2\frac{3}{4} + 1\frac{2}{3} =$ ____ **4)** $5\frac{1}{2} + \frac{7}{12} + \frac{3}{4} =$ ____

5) Pete had to convert all of his fractions to $\frac{?}{24}$ to help with this addition.
His answer is greater than 2.
What calculation was he doing? Try to find more than one solution.

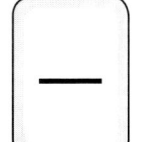 + 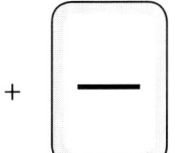 + 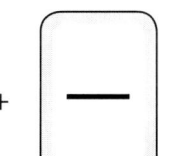 =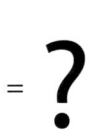

HINT: None of the fractions already had a denominator of 24.

6) Now make up a similar puzzle for a friend or your teacher to solve.

UNIT 10 Subtracting fractions

National Curriculum link:

Add and **subtract fractions with the same denominator and denominators that are multiples of the same number.**

Year 5 pupils should already know that:

- A whole can also be described as a fraction, e.g. $\frac{5}{5}$. This helps us to convert an improper fraction to a mixed number, e.g. $\frac{7}{5}$ to $1\frac{2}{5}$
- It is much easier to use subtraction when fractions have the same denominator

Supporting understanding

Many of the images used for addition can also be used for subtraction. Using knowledge of the inverse, the operation of subtraction will 'undo' addition and vice versa, e.g:

$$\frac{3}{8} + \frac{3}{4} = 1\frac{1}{8} \quad \text{so } 1\frac{1}{8} - \frac{3}{4} = \frac{3}{8}.$$

Children will need to apply knowledge of equivalents to calculate in Year 5:

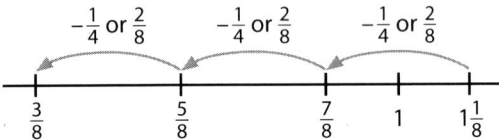

Children may combine the second and third jump as a jump of $\frac{4}{8}$ or $\frac{1}{2}$.

Fraction bars can also be used to support this and help to make sense of the 'find the difference' model of subtraction.

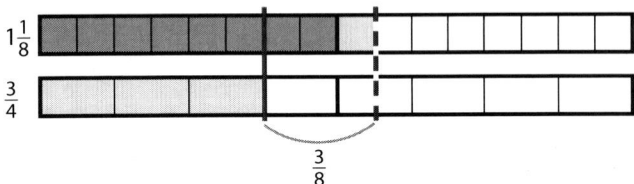

Up to the solid line, both fractions are the same, so the difference of $\frac{3}{8}$ is clearly seen between the solid and the dashed line.

This can also be modelled on the number line.

Recording subtraction calculations

Children in Year 5 will be adding and subtracting fractions that are greater than 1.

Knowledge of improper fraction and mixed number equivalence will play a vital role, e.g:

$$1\frac{1}{8} - \frac{3}{4} = \dots \text{ as } 1\frac{1}{8} - \frac{6}{8} \text{ or } \frac{9}{8} - \frac{6}{8}$$

In the classroom

Begin by counting backwards in steps of $\frac{1}{4}$ starting from $1\frac{7}{8}$.

Ask children to quickly find the first number we land on and any other numbers that will be in the count. Support any groups where needed with reference to fraction bars.

What did you have to do to help with the count?

Establish that it is easier to use the $\frac{1}{4}$ and $\frac{2}{8}$ equivalence so the denominators are the same.

Model the first three steps of the count on a number line.

Consider the next step, $1\frac{1}{8} - \frac{1}{4}$ or $1\frac{1}{8} - \frac{2}{8}$.

What can we do to help us here?

Establish that we can use knowledge of $\frac{8}{8}$ as one whole to help us to bridge the whole.

Ask children also to consider converting to improper fractions, e.g. $\frac{9}{8} - \frac{2}{8} = \frac{7}{8}$.

Record the calculation.

Ask children to use what they know to consider the calculation $1\frac{1}{8} - \frac{3}{4}$. Ask more able children to consider $1\frac{1}{8} - \frac{21}{28}$.

Suggest that the 'finding the difference' model could be useful here as the numbers are in close proximity to each other and to a whole number.

Use the fraction bar image and number line to confirm the difference as $\frac{3}{8}$, also the $\frac{3}{4}$ bar is $\frac{21}{28}$ in its simplest form.

Ask different groups to practise with the following and decide when it may be useful to find the difference:

$$\frac{9}{10} - \frac{4}{5}$$

$$1\frac{2}{3} - \frac{5}{6}$$

$$3\frac{4}{7} - 1\frac{3}{14}$$

Task A 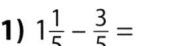 (Independent task or guided learning with an adult)

You will need at least two copies of fraction bars, up to tenths.
Find the answers to the following addition calculations:

1) $1\frac{1}{5} - \frac{3}{5} =$ ____

2) $1\frac{5}{8} - \frac{4}{8} =$ ____

3) $1\frac{2}{3} - \frac{2}{3} =$ ____

4) $1\frac{7}{10} - \frac{4}{10} =$ ____

5) Now make a subtraction calculation of your own using sixths.

Use the fraction bars to help you to find equivalent fractions for the subtractions below.
You may find it useful to 'find the difference'.

6) $1\frac{1}{2} - \frac{3}{4} =$ ____

7) $1\frac{1}{2} - \frac{7}{10} =$ ____

8) $1\frac{3}{4} - \frac{7}{8} =$ ____

Task B (Independent task)

You may need at least two copies of fraction bars, up to twelfths.
Find the answers to the following subtraction calculations.
Look out for questions where it is more useful to 'find the difference'.

1) $1\frac{4}{5} - \frac{3}{10} =$ ____

2) $1\frac{5}{8} - \frac{3}{4} =$ ____

3) $\frac{5}{3} - \frac{5}{6} =$ ____

4) $2\frac{1}{2} - \frac{7}{10} =$ ____

5) Now make a subtraction calculation of your own using sixths and twelfths.

6) Complete the sequence and write the rule.
$\frac{19}{10}, 1\frac{3}{5}, \frac{13}{10},$ ____, ____, ____, ____

> **HINT:** Finding the difference may help you here.

Task C (Independent task)

Find the answers to the following subtraction calculations.
Look out for questions where it is more useful to 'find the difference'.

1) $1\frac{5}{9} - \frac{4}{3} =$ ____

2) $2\frac{4}{9} - \frac{5}{6} =$ ____

3) $2\frac{3}{4} - 1\frac{2}{3} =$ ____

4) $5\frac{1}{2} - \frac{9}{12} - \frac{3}{4} =$ ____

5) Now make a subtraction calculation of your own using sixths and twelfths.
The answer must also be able to be expressed in quarters.

6) Pete and Ami are discussing the 'descending' sequences they have been using.
Find the answers to their questions.

> I have to find the difference between the first number $2\frac{1}{7}$ and the next number in the sequence. The difference is $\frac{5}{14}$. What are the first four numbers in my sequence?

> I have to find the difference between the second number $5\frac{1}{6}$ and the third number in the sequence. The difference is $\frac{4}{3}$. What are the first four numbers in my sequence?

UNIT 11

Counting in decimal steps

National Curriculum link:

[Non-statutory guidance] **Extend counting from year 4, using decimals and fractions including bridging zero, for example on a number line.**

Year 5 pupils should already know that:

- Decimals and fractions are different ways of expressing numbers and proportions
- There is an infinite range of numbers that sit between two whole numbers on a number line
- The position of the digits in a decimal number represents its value

Supporting understanding

Counting on and back in decimal steps will help to prepare children for calculating with decimals.

Place value charts support counting, as children can see how each number is formed, and secure understanding of the value of each digit.

1	②2	3	4	5	6	7	8	9
0.1	0.2	0.3	⊙0.4	0.5	0.6	0.7	0.8	0.9
0.01	0.02	0.03	0.04	⊙0.05	0.06	0.07	0.08	0.09
0.001	0.002	0.003	0.004	0.005	0.006	0.007	0.008	0.009

The numbers circled are 2 ones, 4 tenths and 5 hundredths, which combine to equal 2.45.

Children can explore what happens when e.g. we count on in tenths: hundredths remain the same, tenths increase by one each time and ones will also increase once 10 tenths are counted.

Counting in decimal numbers

Children count across zero in whole numbers in Year 5, but the non-statutory guidance also extends this to decimals and fractions. They have been included in the counting activity here.

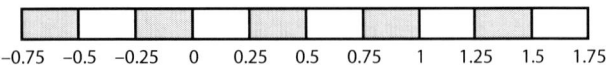

| −0.75 | −0.5 | −0.25 | 0 | 0.25 | 0.5 | 0.75 | 1 | 1.25 | 1.5 | 1.75 |

Here, asking children to predict the value of the last position on a subsequent counting stick requires them to reason about the value of a whole stick (2.5) and then add on to 1.75 to reach 4.25.

We can make links to help children see the equivalence between fraction and decimal representations of the same number, in this case fifths (or tenths):

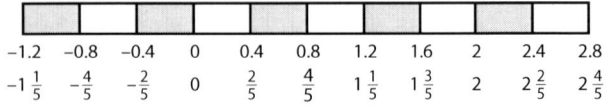

| −1.2 | −0.8 | −0.4 | 0 | 0.4 | 0.8 | 1.2 | 1.6 | 2 | 2.4 | 2.8 |
| $-1\frac{1}{5}$ | $-\frac{4}{5}$ | $-\frac{2}{5}$ | 0 | $\frac{2}{5}$ | $\frac{4}{5}$ | $1\frac{1}{5}$ | $1\frac{3}{5}$ | 2 | $2\frac{2}{5}$ | $2\frac{4}{5}$ |

In the classroom

Explore the structure of the place value chart and the relationship between rows: numbers in the first row are 10 times larger than those in the second row, but 100 times larger than those in the third, etc.

Note the position of the decimal point as we move from, say, 2 to 0.2 to 0.02 and then to 0.002, i.e. it does not move, while the digit 2 moves one place to the right each time.

Ask different groups to consider the numbers that would be in the following counts and how this count would be represented on the place value chart (left):

Start at 2.4 and count on in steps of 0.2

Start at 2.45 and count on in steps of 0.2

Start at 2.45 and count on in steps of 0.15

Make desk-top versions of the chart available.

Introduce the numbers in two different 'mystery' counts:

a) 0.75, 1, 1.25, 1.5, 1.75

b) 1.36, 1.45, 1.54, 1.63

What steps am I counting in? How do you know?

Label a counting stick or number line (10 sections) with the five numbers in mystery count (a), placing 1.75 at the end (see left).

Ask children to find the missing values on the stick. Ask those needing further consolidation to focus on the three positions immediately preceding 0.75.

Explore negative numbers and include the fractional equivalents of $\frac{1}{4}$, $\frac{1}{2}$, $\frac{3}{4}$ on the stick. Also consider mixed number and improper fraction equivalents.

Encourage children to visualise an additional stick at either end and identify the last number on it.

ANSWERS

Task A: Children play Snakes and Ladders

Task B: Children play Snakes and Ladders

Task C: 1) Pete: 2 and 6 or 3 and 4; Ali: two 4s **2)** Ali: he must have rolled two 3s to move three steps of 0.03; Pete cannot move 0.13 in one go

3) Needs to reach 3.53 as it is a different ladder; in two turns as six steps of 0.06 and then five steps of 0.01 or one step of 0.05 **4)** 4 and 2 **5)** E.g. roll two 3s to move 0.09 to get to 3.81, then roll 1 and 6 to move 0.06 to get to 3.87, finally roll 2 and 3 to move 0.06 to get to 3.93

Task A (Independent task)

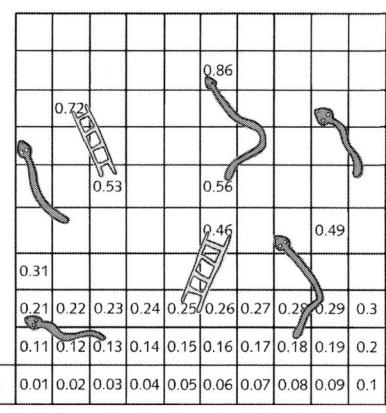

- Take it in turns to roll a number to make a decimal, e.g. 3 becomes 0.03.
- Decide to move your counter on one step or two steps, e.g. one step of 0.03 or two steps of 0.03.
- Ladders will help you to move on quickly, but watch out for the snakes!
- Fill in missing numbers on the grid as you go.
- Write down the counts you make each time, e.g. 0, 0.03, 0.06.
- The winner is the first to reach any number on the top row with their counter.

Task B (Independent task)

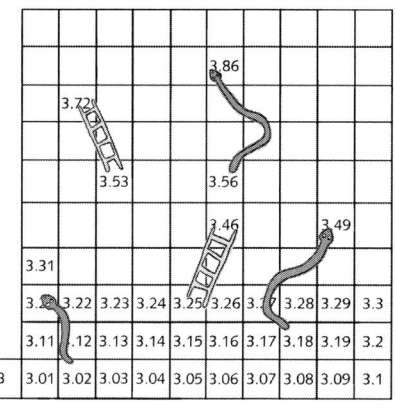

- Take it in turns to roll two dice and move your counter.
- You can decide which number to use as the step size, e.g. 3 becomes 0.03 or 2 becomes 0.02. The other number tells you how many steps to take, e.g. three steps or two steps.
- Ladders will help you to move on quickly, but watch out for the snakes!
- Fill in missing numbers on the grid as you go.
- Write down the counts you make each time, e.g. 3, 3.02, 3.04, 3.06.
- The winner is the first to reach any number on the top row with their counter.

Task C (Independent task or guided learning with an adult)

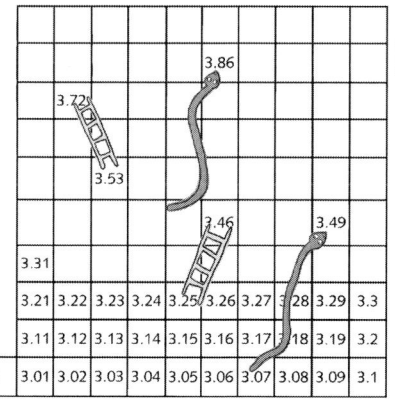

1) Pete and Ali both start on 3. Pete lands on 3.12 on his first go and Ali lands on 3.16.

   Which numbers did they each roll?

   The game continues. They each go up a different ladder. Find out:

2) Who went up the first ladder on their second turn and the numbers they rolled.

3) From 3.12, what is the least number of turns Pete can take to reach a ladder? Describe each count and the numbers rolled.

4) If Ali goes down the second snake, what must he roll to get to 3.53 on his next turn?

5) Pete is on 3.72. He wins the game after three more turns by landing on 3.93. Find a way to make this true, but don't forget the snake!

Pete and Ali take it in turns to roll two dice. They decide which number will be the step size, e.g. 2 becomes 0.02. The other is the number of steps, e.g. use the 3 to count on three steps of 0.02.

UNIT 12 Adding and subtracting decimals

National Curriculum link:

[Non-statutory guidance] **Add and subtract decimals, including a mix of whole numbers and decimals, decimals with different numbers of decimal places, and complements of 1 (for example, 0.83 + 0.17 = 1).**

Year 5 pupils should already know that:

- Decimals and fractions are both ways of representing numbers and proportions
- All fractions can be represented as decimals
- We can use division to help us to convert fractions to decimals, e.g. $\frac{3}{5}$ as $3 \div 5 = 0.6$

Supporting understanding

Children should continue to refine their skills in partitioning in different ways and using number bonds to help to bridge boundaries when adding and subtracting.

By the end of Key Stage 1, children already relate addition and subtraction facts to 10 to facts involving multiples of 10 through use of place value, e.g. using 5 + 3 or 8 – 3 to support 50 + 30 or 80 – 30.

In the same way, they later use the same fact to support the calculations 0.5 + 0.3 or 0.8 – 0.3 and then 0.05 + 0.03 or 0.08 – 0.05.

Counting in decimal steps will also support this.

Measurement also provides plenty of opportunity to calculate with decimals, e.g. 0.8 kg – 0.5 kg and 0.05 m + 0.03 m.

Adding and subtracting decimals

Number bonds also help us to cross boundaries when calculating on a number line, or when adding and subtracting using a column method for calculations that are more difficult to do mentally.

Consider the mental calculation 2.8 + 0.6 = 3.4:

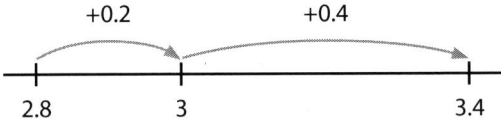

Using knowledge of 28 + 6 and place value will also support here.

Similarly, calculations using decimals to two places can be supported through bonds to 100 and number facts using pairs of two-digit numbers, e.g:

0.83 – 0.45 linked to 83 – 45

1.32 – 0.55 using place value and 55 – 32, to find how much more needs to be subtracted having reached the whole, and then 100 – 23 linked to 1 – 0.23.

ANSWERS

Task A: 1) 1.4 **2)** 1.4 **3)** 0.9 **4)** 0.5 **5)** 9 + 8 = 17 **6)** 1.4 kg

Task B: 1) 1.9 **2)** 0.7 **3)** 0.07 **4)** 0.07 **5)** 3.4 – 0.7 = 2.7 and
5.4 + 0.8 = 6.2 **6)** 0.9 kg

Task C: 1) 0.82 (59 + 23) **2)** 1.59 (80 + 79 or double 80 and subtract 1)
3) 0.48 (97 – 49) **4)** 0.63 (e.g. double 63) **5)** 3.28 + 1.8 = 3.46
6) E.g. Lori 2.1 m and Max 2.86 m; and Lori 3.4 m and Max 2.64 m

In the classroom

Pose this problem for children to consider:

Izzy thinks she can use the same number fact to help her with all of these calculations:

0.8 + 0.5   1.3 – 0.8   0.13 – 0.05

2.8 + 1.5   3.3 – 1.8

What do you think? What did she do?

(Give the calculation 8 + 5 to children who need additional support and ask them to discuss how the fact is used each time.)

Ask children to feed back and look at the use of place value each time. It will be useful to look at other methods that children may have used, e.g. 0.8 + 0.5 as double 0.5 add 0.3, and 3.3 – 1.8 as 3.3 – 1.3 – 0.5.

Model one or two calculations on the number line, drawing attention to the use of number bonds.

Children should choose pairs of decimals from the grid below to add and then subtract using the inverse. They should note the number fact they are using each time. The blank square can be used for numbers of their choice.

0.4	4.8	0.7
3.9	$\frac{4}{5}$	2.5
1.6	0.9	

Ask more able children to extend the use of place value to create numbers such as 0.04 and 0.48 or even 0.004 and 0.048 to make their own grids.

Remind children of the key fact that all fractions can be represented as decimals.

Presenting the two calculations 3.9 kg + 2.5 kg and 0.48 m – 0.25 m, establish that working with decimals helps us to solve problems about measurement.

Task **A** (Independent task)

Use this number fact to help to complete the calculations:

$9 + 5 =$

1) $0.9 + 0.5 =$ ___

2) $0.5 + 0.9 =$ ___

3) $1.4 - 0.5 =$ ___

4) $1.4 - 0.9 =$ ___

5) What number fact did Sami use to help with these calculations?

$1.7 - 0.8 =$ ___ and $0.9 + 0.8 =$ ___

6) Baker Bonny is weighing ingredients for her famous cake.
She puts 0.8 kg of flour and 0.6 kg of fruit into a large bowl.
How much do the ingredients in the bowl weigh in total so far?

HINT: Think about a number fact to help you or show the calculation on a number line.

Task **B** (Independent task)

Use this number fact to help to complete the calculations:

$12 + 7 =$

1) $1.2 + 0.7 =$ ___

2) $1.9 - 1.2 =$ ___

3) $0.12 +$ ___ $= 0.19$

4) $0.19 - 0.12 =$ ___

5) Sami drew this number line to help with his calculations. What are the calculations?

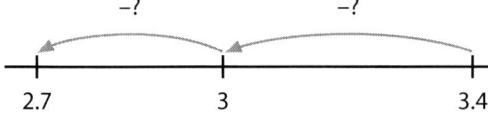
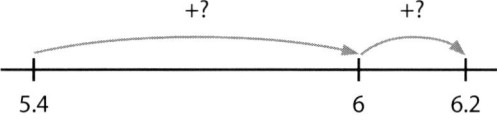

6) Pasta Paulo is weighing ingredients for his famous dish, to be served tonight in his restaurant.
He puts 2.8 kg of cooked pasta and 0.7 kg of cheese into a large bowl.
He then adds some onions, mushrooms and peppers.
The ingredients in the bowl so far weigh 4.4 kg.
How much did the onions, mushrooms and peppers weigh altogether?

Task **C** (Independent task)

Izzy thinks she can use number facts and bonds to 100 to help with each of these calculations.
Complete the calculations and write the fact she used to help her each time:

1) $0.59 + 0.23 =$ ___

2) $0.8 + 0.79 =$ ___

3) $0.97 - 0.49 =$ ___

4) $1.26 - 0.63 =$ ___

5) Izzy drew this number line to help with a different calculation. What is the calculation?

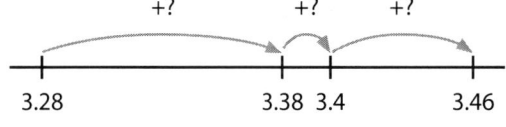

6) Lori and Max are competing in the long jump. They both jump further than 2 m. The difference between the length of their two jumps is 0.76 m. Lori's jump can be written to one decimal place. How far did Lori and Max jump? Find two different solutions.

UNIT 13 Multiplying proper fractions and mixed numbers

National Curriculum link:

Multiply proper fractions and mixed numbers by whole numbers, supported by materials and diagrams.
[Non-statutory guidance] **Connect multiplication by a fraction to using fractions as operators (fractions of), and to division.**

Year 5 pupils should already know that:

- A fraction such as $\frac{3}{4}$ is called a proper fraction, whereas $\frac{4}{3}$ is an improper fraction and $1\frac{1}{3}$ is its mixed number equivalent
- Multiplication and division help us to work with fractions

Supporting understanding

The counting that children have been doing will support them with multiplication of fractions.

This builds on the repeated addition model of multiplication.

Fraction bars, number lines and other practical aparatus will continue to support here, e.g:

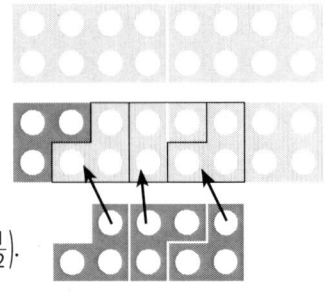

Each whole is made of $\frac{8}{8}$. When we count in steps of $\frac{3}{8}$, the following fractions are made: $\frac{3}{8}, \frac{6}{8}, \frac{9}{8}\left(1\frac{1}{8}\right), \frac{12}{8}\left(1\frac{4}{8} \text{ or } 1\frac{1}{2}\right)$.

So $\frac{3}{8} + \frac{3}{8} + \frac{3}{8} + \frac{3}{8} = 1\frac{1}{2}$.

Therefore, $\frac{3}{8} \times 4 = 1\frac{1}{2}$, or $\frac{3}{8}$ doubled and doubled again.

Recording multiplication calculations

We can use images, as above, and others based on arrays to support the following calculation:

$\frac{3}{4} \times 3 = \frac{9}{4}$ or $2\frac{1}{4}$

In previous years, children will already have met fractions as operators, so this calculation can also be described as $\frac{3}{4}$ of 3.

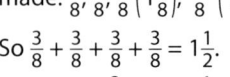
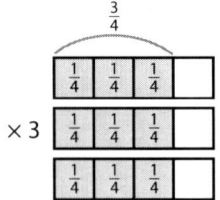

Therefore, the image can also be extended to show $1\frac{3}{4} \times 3$ by partitioning into $1 + \frac{3}{4}$.

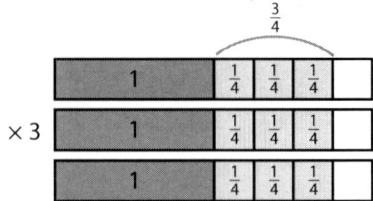

In the classroom

Show the count $\frac{3}{8}, \frac{6}{8}, \frac{9}{8}, \frac{12}{8}, \ldots, \ldots$

What comes next? How do you know?

You may want to suggest to children who need further support that the count could be in steps of either $\frac{2}{8}$ or $\frac{3}{8}$.

Ask children to check and then find the next number in the sequence.

Confirm that the last missing number in the count is $\frac{18}{8}$ or $2\frac{2}{8}\left(2\frac{1}{4}\right)$, and that this is six steps or jumps of $\frac{3}{8}$.

Record as a repeated addition and also as $\frac{3}{8} \times 6$.

Images can also be used to model this, e.g. an extension of the model shown to the left or jumps on a number line.

Now consider the calculation $\frac{3}{4} \times 3$ and use the array-based image or a similar model to prove that the product is $\frac{9}{4}$ or $2\frac{1}{4}$. Look at the use of fractions as operators and revisit the term 'of'.

So $\frac{3}{4} \times 3 = \frac{9}{4}$ or $\frac{3}{4}$ of $3 = \frac{9}{4}$ or $2\frac{1}{4}$.

Show how a similar image can also be used to prove that the product of $1\frac{3}{4} \times 3 = 5\frac{1}{4}$.

By partitioning the calculation into 1×3 and $\frac{3}{4} \times 3$, we can also record this as $(1 \times 3) + \left(\frac{3}{4} \times 3\right) = 3 + 2\frac{1}{4}$.

Invite children to consider how the equivalent improper fraction $\frac{7}{4}$ could also be used here, i.e. $\frac{7}{4} \times 3$, which is $\frac{21}{4}$ or $5\frac{1}{4}$.

Ask different groups to practise with the following, and ask more able children also to consider the use of improper fractions:

$\frac{3}{4} \times 5$ (or $\frac{3}{4}$ of 5) $\frac{5}{6} \times 3$ (or $\frac{5}{6}$ of 3) $1\frac{2}{3} \times 3$ (or $1\frac{2}{3}$ of 3)

ANSWERS

Task A: 1) $\frac{1}{2} \times 5 = 2\frac{1}{2}$ **2)** $\frac{2}{3} \times 4 = \frac{8}{3}\left(2\frac{2}{3}\right)$ **3)** $\frac{3}{5} \times 3 = \frac{9}{5}\left(1\frac{4}{5}\right)$ **4)** $\frac{4}{6}$ or $\frac{2}{3}$ **5)** $\frac{18}{10}$ $\left(1\frac{8}{10}\right)$ **6)** $4\frac{1}{2}$ **7)** $1\frac{1}{2}$ litres

Task B: 1) $\frac{4}{5} \times 4 = \frac{16}{5}\left(3\frac{1}{5}\right)$ **2)** $1\frac{2}{3} \times 5 = 8\frac{1}{3}$ **3)** $\frac{20}{6}\left(3\frac{2}{6} \text{ or } 3\frac{1}{3}\right)$ **4)** $9\frac{1}{10}$ **5)** $20\frac{4}{5}$

6) $11\frac{2}{3}$ litres

Task C: 1) $2\frac{5}{6} \times 5 = 14\frac{1}{6}$ **2)** $14\frac{3}{8}$ **3)** $47\frac{1}{2}$ **4)** E.g. $3\frac{4}{9} \times 9 = 31$ **5)** $180\frac{1}{4}$ gallons

6) $12\frac{11}{12}$ hour which is 12 hours and 55 minutes **7)** 12 days

Task **A** (Independent task or guided learning with an adult)

Which calculations do the following images represent? Find the answer each time.

1)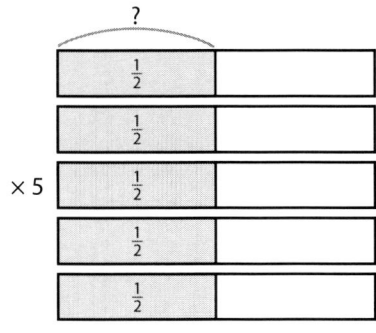

2)

3)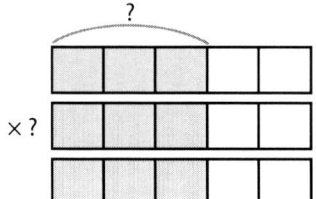

Use fraction bars and counters to help calculate the following:

4) $\frac{1}{6} \times 4 =$ ___

5) $\frac{3}{10} \times 6 =$ ___

6) $1\frac{1}{2} \times 3 =$ ___

7) Izzy uses $\frac{3}{4}$ litre of water a day to water the plants. How many litres does she use in 2 days?

Task **B** (Independent task)

Which calculations do the following images represent? Find the answer each time.

1)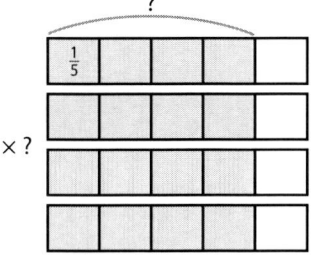

2)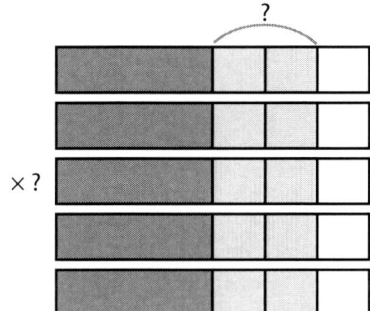

HINT: You can also use fraction bars and counters to help you.

Complete the following calculations:

3) $\frac{5}{6} \times 4 =$ ___

4) $1\frac{3}{10} \times 7 =$ ___

5) $2\frac{3}{5} \times 8 =$ ___

6) Pete drinks $1\frac{2}{3}$ litres of water in a day. How many litres will he drink in a week?

Task **C** (Independent task)

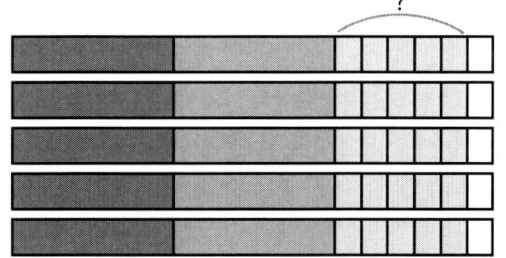

1) Find the answer to the calculation represented by the image.

Complete these calculations.

2) $2\frac{7}{8} \times 5 =$ ___   **3)** $4\frac{3}{4} \times 10 =$ ___

4) $3\frac{4}{9} \times$ ___ $=$ a whole number

5) The delivery truck uses $12\frac{7}{8}$ gallons of fuel in a day.

How many gallons does it use in a fortnight?

Harry spends $2\frac{7}{12}$ of an hour each day on the train.

6) How long does he spend on the train in 5 days? Show the fraction of an hour in minutes.

7) After how many days will he have spent exactly 31 hours in total on the train?

UNIT 14 Problems about multiplying fractions

National Curriculum link:

Multiply proper fractions and mixed numbers by whole numbers, supported by materials and diagrams.

Year 5 pupils should already know that:

- Counting in fraction steps supports the repeated addition model of multiplication
- Partitioning can be used to help to multiply mixed numbers, e.g. $1\frac{2}{3} \times 4$ as 1×4 and $\frac{2}{3} \times 4$
- Converting between improper fractions and mixed numbers may make the calculation easier

Supporting understanding

For children to understand why they are learning the mathematics in the curriculum, it is important that they see how concepts are connected and their purpose in different contexts.

In Year 5, children are expected to 'calculate and compare the area of rectangles (including squares)'. This provides a valuable context to apply multiplication of fractions and to link with arrays, as this underpins understanding of area.

This unit will focus on the area of a rectangle where one length is a whole number.

> I know that the area of this rectangle will be … because …

Area and arrays

An area can be described as 'a measure of the size of any plane surface'.

Consider the rectangle here and how the image used in Unit 13 has been further developed:

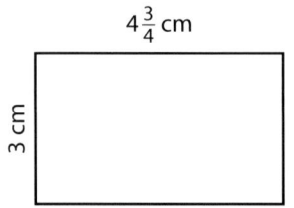

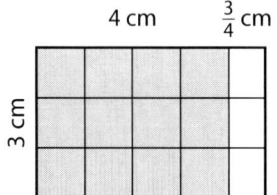

> I know that the area of this rectangle will be less than **15** cm² because the calculation **$4\frac{3}{4} \times 3$** is less than **5×3**.

The area can be calculated as $4\,\text{cm} \times 3\,\text{cm}$ and $\frac{3}{4}\,\text{cm} \times 3\,\text{cm}$. The total area is $14\frac{1}{4}\,\text{cm}^2$.

ANSWERS

Task A: 1) A) More than 6 m² **B)** More than 12 m² **C)** More than 8 m² **2) A)** $7\frac{1}{2}$ m² **B)** 13 m² **C)** 9 m² **3)** $5\frac{1}{2}$ m²

Task B: 1) A) More than 24 m² **B)** More than 15 m² **C)** Less than 16 m² (C) **2) A)** $25\frac{1}{3}$ m² **B)** $16\frac{7}{8}$ m² **C)** $15\frac{1}{5}$ m² **3)** E.g. $19\frac{3}{4}$ m × 4 m and $9\frac{7}{8}$ m × 8 m

Task C: 1) A) Less than 20 m² **B)** More than 8 m² **C)** Less than 25 m² **2) A)** $19\frac{3}{8}$ m² **B)** $9\frac{8}{20}$ m² or $9\frac{2}{5}$ m² **C)** $23\frac{3}{4}$ m² or 23.75 m² **3)** $4\frac{14}{20}$ $\left(4\frac{7}{10}\right)$ m × 4 m or $2\frac{7}{20}$ m × 8 m **4)** E.g. $12\frac{1}{4} \times 6$ m = $73\frac{1}{2}$ m², or $6\frac{1}{8} \times 12$ m = $73\frac{1}{2}$ m² or $10\frac{1}{3} \times 7$ m = $72\frac{1}{3}$ m²

In the classroom

> Revisit some of the calculations that children have been solving recently and the methods used, e.g. $\frac{3}{4} \times 5$, $\frac{5}{6} \times 3$ and $1\frac{2}{3} \times 3$.

> Discuss how estimates are useful when calculating as these help to avoid errors, e.g. a useful estimate of $4\frac{3}{4} \times 3$ is 5×3 (refer to rules of rounding).
>
> *Will the answer be more or less than 15? How do you know?*
>
> Ask more able children to give an example of a calculation where the estimate is still 15, but the answer will be more than 15, e.g. $5\frac{1}{4} \times 3$.

> Introduce the context of area that will be developed during the lesson.
>
> Ask children to discuss the connection between area and multiplication.
>
> Confirm that a rectangle that has dimensions of 5 cm and 3 cm has an area of 15 cm².

> Using the rectangle shown on the left, ask children to refer back to the original estimate they made earlier.
>
> Use the language structure on the left to support.

> Use the second image on the left to model how partitioning the rectangle into wholes and fractions can help us to work out the area of each piece.
>
> Model the calculation to confirm that the area ($14\frac{1}{4}$ cm²) is less than the estimate of 15 cm², as predicted.

> Introduce the problem:
>
> *Symi's family want to rent a house. The property information shows the dimensions of the kitchens.*
>
> *Which house has the largest kitchen? Make an estimate first.*
>
> A) $3\frac{1}{4}$ m × 4 m B) $1\frac{9}{10}$ m × 6 m C) $3\frac{5}{8}$ m × 3 m
>
> *What do you notice about each of the estimates?*
>
> *Will the area be more or less than the estimate?*
>
> Ask different groups to work on each house (house A being the least challenging).

Task **A** (Independent task or guided learning with an adult)

Izzy and her family are also looking at houses.
They compare the size of each bedroom.
Here are the three bedrooms in one of the houses.

1) Estimate the area of each bedroom. Use the speech bubble to help you.

> I know that the area of this rectangle will be ___ because ___ .

2) Calculate the area of each bedroom.

3) Izzy measures the bathroom as $2\frac{3}{4}$m × 2 m. What is the area?

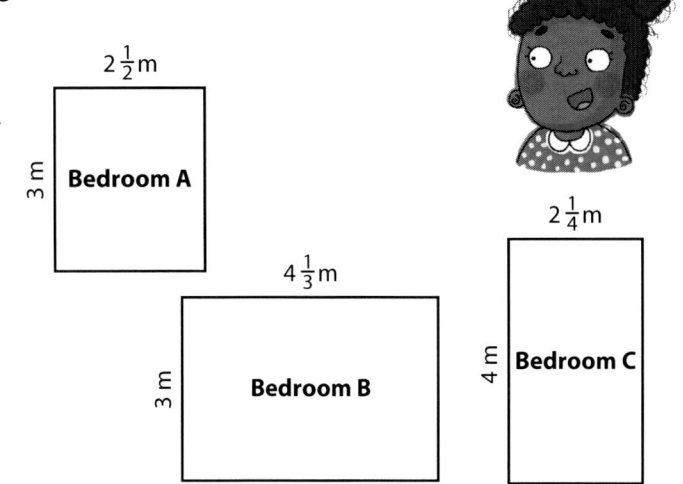

Task **B** (Independent task)

Symi's family want to compare the size of the bedrooms in one of the houses.

1) Estimate the area of each bedroom. Use the speech bubble to help you.

> I know that the area of this rectangle will be ___ because ___ .

2) Calculate the area of each bedroom.

3) Symi estimates that the garden has an area of 80 m². The actual area is less than this.

   What are the possible dimensions of the garden?

   Find more than one solution.

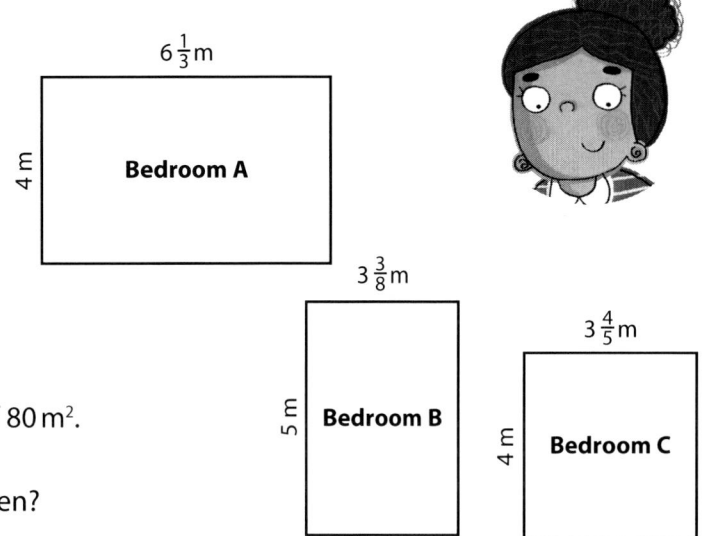

Task **C** (Independent task)

Pete and his family are also looking at houses. They compare the size of each bedroom.

Here are the dimensions of three bedrooms in one of the houses:

Bedroom A	Bedroom B	Bedroom C
$3\frac{7}{8}$m × 5 m	$2\frac{7}{20}$m × 4 m	4.75 m × 5 m

> I know that the area of this rectangle will be ___ because ___ .

1) Estimate the area of each bedroom. Use the speech bubble to help you.

2) Calculate the area of each bedroom.

3) The area of the garage is exactly double the area of Bedroom B.

   Suggest the possible dimensions of the garage.

4) Pete estimates that the garden has an area of 72 m². The actual area is a little more than this, but not as much as 74 m². What are the possible dimensions of the garden?

   Find at least two different solutions.

UNIT 15 Fractions of amounts and remainders

National Curriculum link:

[Non-statutory guidance] **Interpret non-integer answers to division by expressing results in different ways according to the context, including with remainders, as fractions, as decimals or by rounding (for example, $98 \div 4 = \frac{98}{4} = 24 \, r \, 2 = 24\frac{1}{2} = 24.5 \approx 25$).**

[Non-statutory guidance] **Connect multiplication by a fraction to using fractions as operators (fractions of), and to division.**

Year 5 pupils should already know that:

- A multiplication such as $\frac{3}{4} \times 5$ can also be written as $\frac{3}{4}$ of 5 (fractions as operators) and vice versa
- A fraction such as $\frac{5}{8}$ can be re-written as $5 \div 8$ using knowledge of division, and vice versa

Supporting understanding

A fraction such as $\frac{5}{8}$ can be re-written as $5 \div 8$ using knowledge of division, and vice versa.

From Key Stage 1, children will have been relating finding unit fractions of amounts to division, e.g. $\frac{1}{8}$ of 32 can also be found by dividing 32 by 8.

In Year 4, children 'understand the relation between non-unit fractions and multiplication and division of quantities'.

Images used in previous years will continue to support children to find a unit fraction of an amount then the non-unit fraction of the same amount, e.g. $\frac{5}{8}$ of 32 by first finding $\frac{1}{8}$.

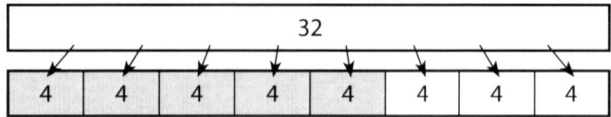

Children in Year 5 will now know that finding $\frac{5}{8} \times 32$ will result in the same answer.

Fractions as remainders

Consider the calculation $38 \div 8$ or $\frac{1}{8}$ of 38. By partitioning 38 into a multiple of 8 and the remaining amount we can calculate this more easily, i.e. as $32 \div 8$ and $6 \div 8$. Since we already know that $6 \div 8$ can be written as $\frac{6}{8}$, the answer to the original calculation is $4\frac{6}{8}$ or $4\frac{3}{4}$ or 4.75.

We may well have recorded the answer as 4 r 6 but we need to know the value of the remainder, especially in work on measurement or on finding the mean as an average (in Year 6). Knowing the value of $\frac{1}{8}$ of 38 will help us to find $\frac{3}{8}$ by multiplying $4\frac{3}{4}$ or 4.75 by 3.

ANSWERS

Task A: Discussion and game

Task B: E.g. $\frac{1}{3}$ of 34 litres = $4\frac{1}{3}$ litres so $\frac{2}{3}$ of 34 litres = $8\frac{2}{3}$; $\frac{1}{8}$ of 34 litres = $4\frac{1}{4}$ litres so $2\frac{1}{8}\left(\frac{1}{4}\right)$ of 34 litres = $8\frac{1}{2}$ litres

Task C: E.g. $\frac{1}{3}$ of 79 litres = $26\frac{1}{3}$ litres so $\frac{5}{3}$ of 79 litres is $131\frac{2}{3}$ litres; $\frac{1}{8}$ of 79 litres = $9\frac{7}{8}$ litres so $2\frac{1}{8}\left(\frac{1}{4}\right)$ of 79 litres = $19\frac{6}{8}$ or $19\frac{3}{4}$ or 19.75 litres

In the classroom

List these, or similar, calculations and ask children to sort them into those that are easier to solve and those that are more difficult:

$\frac{1}{8}$ of 32 $38 \div 8$ $75 \div 10$ $75 \div 9$

Ask children to suggest other calculations that could be sorted in this way.

Ask children to feed back. Establish that when the divisor or denominator is a factor of the number to be divided, there will be no remainders. Look at the use of place value to support $75 \div 10$.

Return to $38 \div 8$. Consider the remainder after division.

Use the model on the left to partition 38, then to express the remainder as a fraction by making the link to the key point that $\frac{6}{8}$ can be written $6 \div 8$.

Use equivalent fractions of $\frac{6}{8}$ and $\frac{3}{4}$ to express the remainder in the simplest form and consider the decimal equivalent.

Ask children to discuss how they can now use this knowledge to help them to find $\frac{3}{8}$ of 38 or $\frac{5}{8}$ of 38.

Ask more able children: '*How about $\frac{3}{7}$ of 38?*'

Ask groups to use one or two dice to generate a fraction, e.g. roll 4 to create $\frac{1}{4}$ or use two dice to roll a 3 and 5 to create $\frac{3}{5}$. (More able children may also wish to consider $\frac{5}{3}$.) They should choose to find this fraction of either or both of these numbers:

24   45

Challenge children to write the remainder as a fraction in its simplest form and as a decimal.

List the different fractions found and consider any decimal equivalents. Remind children that, e.g:

$\frac{2}{3}$ of 45 can also be written as $\frac{2}{3} \times 45$.

Task **A** (Guided learning with an adult)

You will need a multiplication square and dice. This session will focus on finding non-unit fractions of amounts to build on the unit fractions found by this group in the main teaching.

- Ask children to look for the multiplication tables where the number 32 can be found.

  How would this help us to find the answer to 32 ÷ 4?

- Pose the problem that we need to find 32 ÷ 5 or $\frac{1}{5}$ of 32.

  Why is this more difficult?

- Check the multiplication table to confirm that 32 is not a multiple of 5 and refer to what we know about the ones digit in a number that is a multiple of 5.

- Discuss and invite children to have a go at the problem using what they found out in the main lesson about partitioning and remainders as fractions.

  How does this help us to find $\frac{2}{5}$ or $\frac{3}{5}$ of 32?

Now play the dice game as a group, using two dice to generate non-unit fractions.

Task **B** (Independent task)

34 litres

Play the dice game to find different fractions of 34 litres.

When you roll a 4 or a 5, you can choose to double numbers, so you can make eighths or tenths as well as quarters and fifths, i.e. you can make $\frac{4}{5}$, $\frac{4}{10}$, $\frac{5}{8}$ or $\frac{8}{10}$ here!

Find a way to record all the fractions you find.

Remember to try to write your fraction remainders in their simplest form and as any decimals you know.

Task **C** (Independent task)

79 litres

Play the dice game to find different fractions of 79 litres.

HINT: Remember that improper fractions can also be written as mixed numbers, e.g. $\frac{5}{3}$ as $1\frac{2}{3}$.

When you roll a 4 or a 5, you can choose to double numbers, so you can make eighths or tenths as well as quarters and fifths, i.e. here you can make $\frac{4}{5}$, $\frac{4}{10}$, $\frac{5}{8}$ or $\frac{8}{10}$!

Find a way to record all the fractions you find.

Remember to try to write your fraction remainders in their simplest form and as any decimals you know.

Challenge yourself to think about making improper fractions with the dice.

UNIT 16 Rounding decimals

National Curriculum link:

Round decimals with two decimal places to the nearest whole number and to one decimal place.

Year 5 pupils should already know that:

- Decimals and fractions are different ways of expressing numbers and proportions
- There is an infinite range of numbers that sit between two whole numbers on a number line
- The position of the digits in a decimal number represents its value

Supporting understanding

Previously, the blank 100 square image and base 10 equipment have been used to support understanding of decimals.

It is also easy to see the link between decimals as fractions, e.g. $\frac{1}{4}$ or $\frac{25}{100}$, which will also support work on percentages.

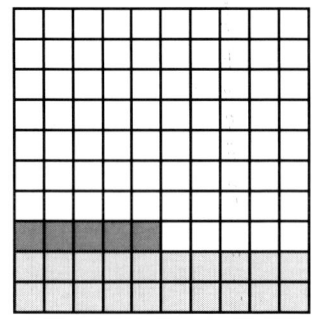

We can also see that $\frac{25}{100}$ or 0.25 is not as large as a half or 0.5 and that, using the rules of rounding, it rounds down to zero.

This can also be confirmed on a number line:

0.25 rounds to **0** because it only has **2 tenths** so it is closer to **0** than **1**.

Rounding decimals

Using the rule for rounding to the nearest ten, we can also round to the nearest tenth or whole number. Rounding to the nearest hundred can also help us to round to the nearest whole number if we look at the number of hundredths.

  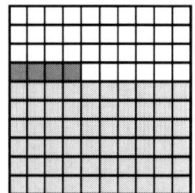

In this case, 2.64 has 6 tenths (or 64 hundredths), which is more than 5 tenths, so we must round up to the nearest whole number (3).

As there are only 4 hundredths in 2.64, it rounds down to the nearest tenth (2.6).

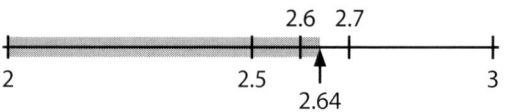

In the classroom

Display a number line from 0 to 5. Ask different groups to decide and be prepared to explain where these pairs of numbers should be placed:

1.5 and 2.75   0.6 and 3.99   4.32 and 5.07

Take feedback and encourage children to improve explanations using language such as 'tenths', 'hundredths', 'between', 'close to', 'more than', 'less than', 'equal to'.

Look for children who have also made links to fractions, i.e. $\frac{1}{2}$, $\frac{3}{4}$ and $\frac{3}{5}$, to help them.

Ask more able children to explain why 5.07 does not sit on this number line unless it is extended past 5.

Ask children to focus on their largest number. They are all to two decimal places.

Which two whole numbers is your decimal number between? Is it more or less than halfway between the whole numbers? How does this help us to round the decimals to the nearest whole number?

Refer to the rules of rounding and use the language structure to support understanding.

Present each number using the base 10 images to confirm the decisions made and to support rounding to the nearest tenth. Also refer back to the number line here and ask children to consider other numbers that will round to the same tenth.

Being able to round decimals to the nearest whole number or tenth will help us to make estimates when calculating and with measurement, e.g. to round 1.68 m to the nearest tenth of a metre (10 cm) and then to the nearest whole metre.

ANSWERS

Task A: 1) 2 **2)** 3 **3)** 1 **4)** 6 **5)** 5 **6)** 4 m **7)** 5 m **8)** 4 m **9)** 3.6 m, 5.5 m, 4.3 m

Task B: 1) 2 **2)** 2 **3)** 10 **4)** 6 **5)** 5 **6)** 3.5 m and 4 m **7)** 5 m and 5 m
8) 4.3 m and 4 m **9)** 10 m and 10 m **10)** E.g. 2.95 m and 3.01 m

Task C: 1) 13.5 m and 14 m **2)** 5 m and 5 m **3)** 3.5 m and 3 m
4) 9.9 m and 10 m **5)** 10.1 m and 10 m **6)** Range of original length is 2.51 m to 2.59 m so that when 10 cm is cut off the range becomes 2.41 m to 2.49 m, which rounds down to 2 m **7)** When 10 cm is cut off it is 2.51 m and this still rounds up to 3 m **8)** Rounds down to 3 kg as there are only 4 tenths of a kg

Task **A** (Independent task)

Round each of these numbers to the **nearest whole number**.
Use the speech bubble to help you to explain.

___ rounds to ___ because ___ .

1)

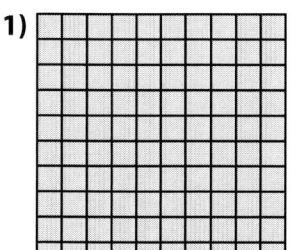 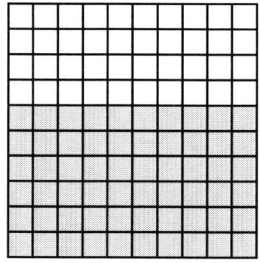

2) 2.65   **3)** 1.48   **4)** 6.25   **5)** 4.9

Now round each of these lengths to the **nearest metre**.

6) 3.64 m   **7)** 5.48 m   **8)** 4.25 m

9) Now round each of these same lengths to the **nearest tenth of a metre**.

Task **B** (Independent task)

Round each of these numbers to the nearest whole number.
Use the speech bubble to help you to explain.

1) 2.36   **2)** 1.65   **3)** 10.48
4) 6.05   **5)** 4.83

___ rounds to ___ because ___ .

Now round each of these lengths to the nearest tenth of a metre and then the nearest metre.

6) 3.51 m   **7)** 5.03 m
8) 4.25 m   **9)** 9.95 m

10) Ahmed did the same with two different lengths. They both rounded to 3 metres and to the same tenth of a metre, but none of the digits in either number is the same.
Find a way to make this true.

Task **C** (Independent task)

Round each of these lengths to the nearest tenth of a metre and then the nearest metre.

1) 13.51 m   **2)** 5.01 m   **3)** 3.49 m   **4)** 9.93 m   **5)** 10.11 m

6) Izzy has a length of string. She measures it to two decimal places.
The length rounds to 3 m, when rounded to the nearest metre.
She then cuts 10 cm off the length of the string. It no longer rounds to 3 m.
Find a way to explain what has happened and give the range of possible lengths for Izzy's original piece of string.

7) Pete thinks that her original string was 2.61 m. Explain how you know he has made a mistake.

8) How would you round 3.468 kg to the nearest kg?

UNIT 17
Comparing and ordering numbers with up to three decimal places

National Curriculum link:

Read, write, order and compare numbers with up to three decimal places.

Year 5 pupils should already know that:

- Decimals and fractions are different ways of expressing numbers and proportions
- There is an infinite range of numbers that sit between two whole numbers on a number line
- The position of the digits in a decimal number represents its value

Supporting understanding

Ordering and comparing numbers relies on our understanding of place value. If this is not secure with whole numbers, decimals are likely to prove a problem.

Images like the place value chart help to partition and combine numbers and, when used alongside a place value grid, can secure the relationship between e.g. 2, 0.2, 0.02 and 0.002.

1	2	3	4	5	6	7	8	9
0.1	0.2	0.3	0.4	0.5	0.6	0.7	0.8	0.9
0.01	0.02	0.03	0.04	0.05	0.06	0.07	0.08	0.09
0.001	0.002	0.003	0.004	0.005	0.006	0.007	0.008	0.009

We can combine these as $2 + 0.2 + 0.02 + 0.002 = 2.222$:

1	0.1	0.01	0.001	
2				2
0	2			0.2
0	0	2		0.02
0	0	0	2	0.002

Comparing and ordering decimal numbers

In the same way that we have used base 10 equipment for tenths and hundredths, it can also be extended to include thousandths so that everything becomes 10 times smaller. It takes 1000 small cubes to fill the large cube so each is $\frac{1}{1000}$ or 0.001.

1.235 can be represented in this way and can be compared to e.g. 1.245. We know that 1.245 is larger as it has 0.01 more than 1.235.

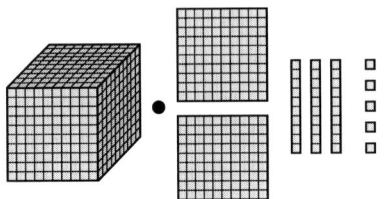

Partitioning will also help to confirm this, e.g:

$1.235 = \cancel{1} + \cancel{0.2} + 0.03 + \cancel{0.005}$

$1.245 = \cancel{1} + \cancel{0.2} + 0.04 + \cancel{0.005}$

Parts that are equal are in this way identified.

In the classroom

Use the place value chart, or similar, to count in different steps, e.g. start at 0.07 and count in steps of 0.01 or 0.02, paying particular attention to the step from 0.09 to 0.1 (or 0.11 if counting in steps of 0.02). Stop at 0.23.
How has this number been made? (0.2 and 0.03)

Ask different groups to quickly find the parts that will be needed for 1.23, 1.235 and 1.245. List the parts needed so links to partitioning can be made:
$1.23 = 1 + 0.2 + 0.03$
$1.235 = 1 + 0.2 + 0.03 + 0.005$
$1.245 = 1 + 0.2 + 0.04 + 0.005$

How can we use what we have found out to help us to place these numbers in order? Which is the largest? How do you know?
Different groups could be in charge of checking the ones and tenths, another group to check the hundredths, etc.
Establish that all numbers have an equal number of ones and tenths, and two of the numbers have an equal number of thousandths, etc.

Agree that 1.23 is the smallest as it has no thousandths and has only 3 hundredths. 1.245 is the largest.
Other images or resources can also be used to confirm this.

Ask different groups to order sets of numbers, using partitioning to help or to prove their decisions:
3.46, 3.36, 3.44
3.456, 3.356, 3.446
3.488, 3.398, 3.46
Ask more able children to include a number of their own that goes between their smallest and middle numbers.

ANSWERS

Task A: 1) 4.26 = 4 + 0.2 + 0.06, etc. **2)** E.g. 3.26 smaller than 4.26 **3)** 3.26, 4.26, 4.27, 4.36 **4)** 5.371 as it has 7 hundredths and 5.361 only has 6 hundredths

Task B: 1) 5.126 = 5 + 0.1 + 0.02 + 0.006, etc. **2)** E.g. 5.123 is smaller than 5.13 **3)** 4.126, 4.926, 5.123, 5.126, 5.13, 5.226 **4)** E.g. 3.379, 3.513, 3.555; 3.355, 3.379, 3.555, where the largest possible number is 3.555

Task C: E.g. 2.35, 2.354, 2.401, 2.413, 2.43, 2.435, 2.513

Task **A** (Independent task)

4.26	3.26	4.36	4.27

1) Use the place value chart to help you to partition each of these numbers.

Record as 4.26 = 4 + ___ .

2) Choose pairs of numbers and find which is smaller each time.

3) Use what you have found out to order all four numbers from the smallest to the largest.

4) Which of these two numbers is the largest?

Find a way to prove your answer.

5.371	5.361

Task **B** (Independent task)

5.126	4.126	5.226	5.123	5.13	4.926

1) Use the place value chart to help you to partition each of these numbers.

Record as 5.126 = 5 + ___ .

2) Choose pairs of numbers and find which is smaller each time.

3) Use what you have found out to order all six numbers from the smallest to the largest.

4) Sami orders three different numbers with three decimal places. All digits are odd.

All numbers have a units (ones) value of 3.

The largest number does not include the digits 7 and 9.

The smallest number does not include the digit 1.

Find a way to make this true.

Task **C** (Independent task or guided learning with an adult)

Ashton orders two numbers with **2 decimal places** and five numbers with **3 decimal places** in this grid.

- They are all between $2\frac{2}{4}$ and $2\frac{3}{5}$ on the number line.
- All numbers have only the digits 0, 1, 2, 3, 4 and 5.
- No digit is repeated in a number, e.g. 2.112 is not possible.
- Two numbers are made up of the same digits, but in a different order.
- The largest number is exactly 0.1 more than the middle number.

Find different ways to make this true.

UNIT 18 Solving problems about numbers with up to three decimal places

National Curriculum link:

Solve problems involving numbers with up to three decimal places.

Year 5 pupils should already know that:

- Decimals and fractions are different ways of expressing numbers and proportions
- There is an infinite range of numbers that sit between two whole numbers on a number line
- The position of the digits in a decimal number represents its value

Supporting understanding

Children should have regular opportunities to make decisions about the position of decimals or fractions on a number line.

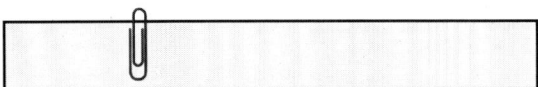

Consider the number line and paper clip shown here and the following questions:

When the start of the line is 0 and the end is 1 kg, what number could the clip represent? (0.25 kg)

When the start is still 0 but the end is $\frac{1}{2}$ kg, why can't the clip represent the same number?

What can it represent now? (0.125 kg)

Where will 0.125 kg sit on the 0 to 1 kg number line? Is it larger or smaller than 0.25 kg?

> I know … so …

> It could be … because …
> It couldn't be … because …

Language structures will continue to support children's reasoning and explanations.

Solving problems about decimal numbers

Money always provides useful contexts to work with numbers with two decimal places and helps to secure the fact that e.g. $\frac{2}{10}$ of a pound or 0.2 is equivalent to 0.20, as this is required for the notation £0.20.

We can easily explore numbers with three decimal places through mass (1 kg = 1000 g), capacity (1 litre = 1000 ml) and length (1 km = 1000 m).

This unit requires children to order, round and calculate with decimals, particularly in relation to complements of 1, e.g. 0.73 + 0.27.

In the classroom

Give children a number line and paper clip as shown (left). (This could also be a fraction bar 'whole' to later link with quarters).

Given that the number line represents 0–1, ask different groups to position the clip at 0.5, 0.25 and 0.75 (0.5 being the least and 0.75 the most challenging, particularly with the following task).

Confirm positions by looking at the value of each digit in the number and linking to the fractions $\frac{1}{2}$, $\frac{1}{4}$ and $\frac{3}{4}$.

Pose the problem that the number line now represents 0 to $\frac{1}{2}$.

What does each paper clip represent now?

What can we use to help us?

Each group should consider their own paper clip.

Link 0.25 to a quarter as half of a half to support.

Link half of $\frac{1}{4}$ to $\frac{1}{8}$ (0.125) and half of $\frac{3}{4}$ to $\frac{3}{8}$ (0.375).

Ask children to check with a calculator.

Reveal a 0 to 0.5 kg scale.

Where would 250 g, 125 g and 375 g go on the scale?

Establish the equivalence between 1 g and $\frac{1}{1000}$ kg or 0.001 kg so 125 g is equal to $\frac{125}{1000}$ kg 0.125 kg.

Return to 0.25 kg. Discuss how much more we need to add to equal 1 kg.

Link to complements to 1 using number bonds.

Now round 0.25 to the nearest tenth of a kg.

How can we round 0.125 kg and 0.375 kg to the nearest tenth of a kg?

ANSWERS

Task A: 1) 2.25 kg **2)** 4.65 kg **3)** 0.75 kg **4)** 0.35 kg **5)** Heavier
Task B: 1) 2.45 kg **2)** 2.625 kg **3)** 0.55 kg **4)** 0.375 kg **5)** 2.45 kg, 2.575 kg, 2.625 kg **6)** 2.5 kg, 2.6 kg, 2.6 kg

Task C: 1) 2.475 kg apples and 2.55 kg bananas **2)** 0.525 kg apples and 0.45 kg bananas **3)** 2.475 kg, 2.525 kg, 2.55 kg **4)** 2.5 kg, 2.5 kg, 2.6 kg
5) To three decimal places, the range is 2.526–2.549 kg

Task **A** (Independent task)

Pete is weighing apples and bananas.

He uses his number line and paper clip to show what happens.

1) How many kilograms (kg) of apples does he have?

2) How many kilograms (kg) of bananas does he have?

3) How many more kilograms of apples does he need to reach 3 kg?

4) How many more kilograms of bananas does he need to reach 5 kg?

5) Pete also weighs 2.3 kg of pears. Are the pears heavier or lighter than the apples?

Apples

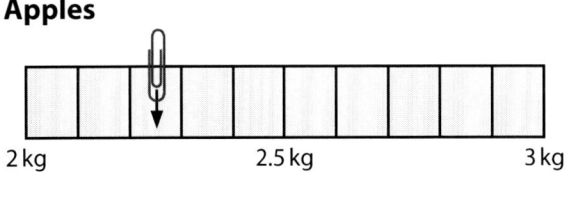

2 kg   2.5 kg   3 kg

Bananas

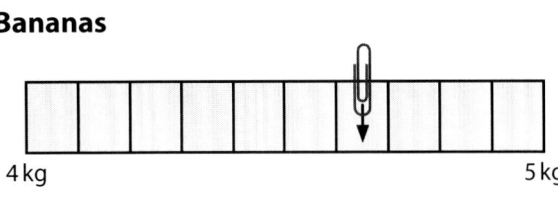

4 kg   5 kg

Task **B** (Independent task or guided learning with an adult)

HINT: Check the scale each time.

Izzy is weighing apples and bananas.

She uses her number line and paper clip to show what happens.

1) How many kilograms (kg) of apples does she have?

2) How many kilograms (kg) of bananas does she have?

3) How many more kilograms of apples does she need to reach 3 kg?

4) How many more kilograms of bananas does she need to reach 3 kg?

5) Izzy also weighs 2.575 kg of pears. Order all the fruits from lightest to heaviest.

6) Now round the mass of each fruit to the nearest tenth of a kilogram.

Apples

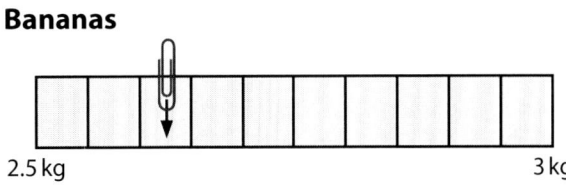

2 kg   2.5 kg

Bananas

2.5 kg   3 kg

Task **C** (Independent task)

HINT: Check the scale each time.

Symi is weighing apples and bananas.

She uses her number line and paper clip to show what happens.

1) How many kilograms (kg) of apples and how many kilograms (kg) of bananas does she have?

2) How many more kilograms of apples does she need to reach 3 kg? And bananas?

3) Symi also weighs 2.525 kg of pears. Order all the fruits from lightest to heaviest.

4) Now round the mass of each fruit to the nearest tenth of a kilogram.

5) Symi also weighed oranges. The oranges are heavier than the pears, but still round to 2.5 kg. What is the range of possibilities for the mass of the oranges?

Apples

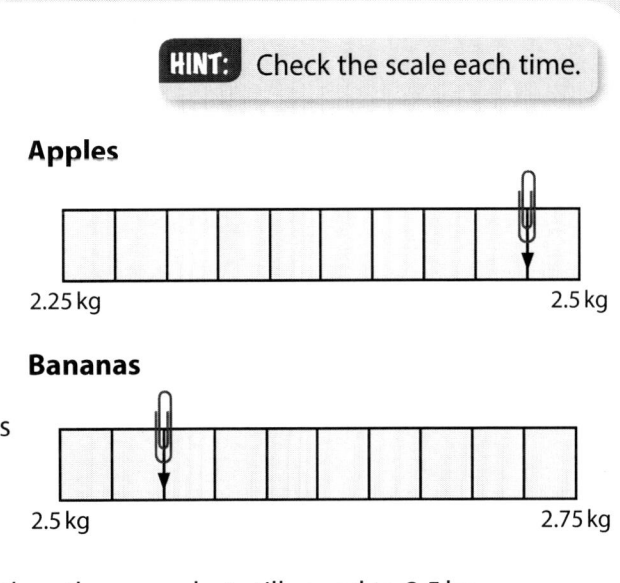

2.25 kg   2.5 kg

Bananas

2.5 kg   2.75 kg

UNIT 19 — Linear sequences involving fractions and decimals

National Curriculum link:

[Non-statutory guidance] **Recognise and describe linear number sequences, including those involving fractions and decimals, and find the term-to-term rule.**

Year 5 pupils should already know that:

- Decimals and fractions are different ways of expressing numbers and proportions
- There is an infinite range of numbers that sit between two whole numbers on a number line
- The position of the digits in a decimal number represents its value

Supporting understanding

In previous units, children have counted on and back in fraction and decimal steps, e.g:

$\frac{3}{10}, \frac{6}{10}, \frac{9}{10}, \frac{12}{10}, \frac{15}{10}$

$1\frac{3}{5}, 1\frac{1}{5}, \frac{4}{5}, \frac{2}{5}, 0$

0.5, 0.8, 1.1, 1.4, 1.7

This will have extended to further activities in the classroom that require children to identify the value of the step, having been given some numbers in the count.

This could include e.g:

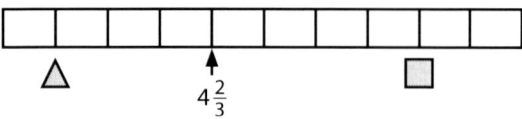

$4\frac{2}{3}$

When the step size is $\frac{2}{3}$, what are the values of the triangle and the square?

When the value of the triangle is $\frac{2}{3}$, what is the step size and what is the value of the square?

Linear sequences

A number pattern that increases (or decreases) by the same amount each time is called a linear sequence.

Children need to know that, to determine the term-to-term rule in a sequence, we need to know at least two terms and the positions of these terms.

So knowing that the terms 0.2 and 0.4 are included could mean any of the following sequences, unless we know their positions:

0.2, 0.4, … , … , …
0.2, … , 0.4, … , …
0.2, … , … , 0.4, …
0.4, 0.2, … , … , …

When we know that the first term is 0.4 and the third is 0.2, this describes a sequence that is not shown here at all.

In the classroom

Begin by setting different groups off on different counts, asking them to record the steps and find the 10th number in the count, e.g:

Starting on 3.75 or $3\frac{3}{4}$ and counting on in steps of 0.25 or $\frac{1}{4}$.

Starting on 10.25 or $10\frac{1}{4}$ and counting back in steps of 0.75 or $\frac{3}{4}$.

Starting on 10.5 and counting back in steps of $\frac{1}{8}$ or 0.125.

Explain that we can call these counts 'linear sequences', as they are number patterns that increase or decrease by the same amount each time. Refer to one of the counts. Explain that the sequence is made up of terms that follow a rule, e.g. on in steps of 0.25 or + 0.25.

But what are the term-to-term rules for these linear sequences?

1.4, 1.1, 0.8, … , …
1.35, 1.2, 1.05, 0.9, … , …
0.675, … , 0.525, … , …

Make calculators available to check or support.

Ask children to feed back on rules and provide missing terms.

Record the rules as subtract 0.3, subtract 0.15 and subtract 0.075.

Show the sequence: … , $3\frac{1}{2}$, … , …

What is the term-to-term rule for this sequence?

Ask children to suggest some possible sequences and the rule each time. Include rules where the sequence crosses zero, e.g. subtract 3.

Establish that we need to know at least two terms in the sequence to be certain of the rule.

ANSWERS

Task A: 1) $10\frac{1}{2}, 9\frac{1}{2}$; subtract 1 **2)** 1 and 1.4; add 0.2 **3)** 5, $4\frac{3}{5}$; subtract $\frac{2}{5}$ **4)** 4.05, 4.15, 4.25; add 0.1 **5)** $5\frac{3}{4}, 8\frac{1}{4}, 10\frac{3}{4}, 13\frac{1}{4}, 15\frac{3}{4}$ **6)** 10.8, 10.3, 9.8, 9.3, 8.8

Task B: 1) $12\frac{4}{8}, 11\frac{7}{8}$; subtract $\frac{5}{8}$ **2)** 0.68 and 0.76; add 0.04 **3)** $7\frac{4}{5}$ and 5; subtract $1\frac{2}{5}$ **4)** 3.32, 3.11; subtract 0.21 **5)** First is $-1\frac{1}{3}$ and then 4, $6\frac{2}{3}, 9\frac{1}{3}$, 12 **6)** First is 13.03 and then 10.99, 9.97, 8.95, 7.93

Task C: 1) $11\frac{6}{7}$, 11; subtract $\frac{6}{7}$ **2)** 3.54 and 3.12, 2.91, 2.7; subtract 0.21 **3)** $9\frac{6}{9}$ and $8\frac{1}{9}, 7\frac{3}{9}$; subtract $\frac{7}{9}$ **4)** 0.098 and 0.126; add 0.014 **5)** $-\frac{6}{7}, 1\frac{5}{7}$ and fifth is $9\frac{3}{7}$ **6)** 14.77, 13.685 and fifth is 10.43 **7)** Yes; rule is add 0.09, related to multiples of 9, where 81 is the tenth term and 9 × 0 is the first

Task **A** (Independent task)

Find the term-to-term rule for each of these sequences.
Find all the missing terms.

1) $13\frac{1}{2}$, $12\frac{1}{2}$, $11\frac{1}{2}$, ___ , ___

2) 0.6, 0.8, ___ , 1.2, ___

3) $5\frac{4}{5}$, $5\frac{2}{5}$, ___ , ___ , $4\frac{1}{5}$

4) 3.75, 3.85, 3.95, ___ , ___ , ___

5) The rule is 'add $2\frac{1}{2}$'. The first term in the sequence is $3\frac{1}{4}$.
Write the next five terms.

6) The rule is 'subtract 0.5'. The first term in the sequence is 11.3.
Write the next five terms.

Task **B** (Guided learning with an adult)

Find the term-to-term rule for each of these sequences.
Find all the missing terms.

1) $14\frac{3}{8}$, $13\frac{6}{8}$, $13\frac{1}{8}$, ___ , ___

2) 0.6, 0.64, ___ , 0.72, ___

3) $9\frac{1}{5}$, ___ , $6\frac{2}{5}$, ___ , $3\frac{3}{5}$

4) 3.95, 3.74, 3.53, ___ , ___

5) The rule is 'add $2\frac{2}{3}$'. The second term in the sequence is $1\frac{1}{3}$.
Find the first term and then write the next four terms after $1\frac{1}{3}$.

6) The rule is 'subtract 1.02'. The second term in the sequence is 12.01.
Find the first term and then write the next four terms after 12.01.

Task **C** (Independent task)

Find the term-to-term rule for each of these sequences.
Find all the missing terms.

1) $14\frac{3}{7}$, $13\frac{4}{7}$, $12\frac{5}{7}$, ___ , ___

2) 3.75, ___ , 3.33, ___ , ___ , ___

3) $10\frac{4}{9}$, ___ , $8\frac{8}{9}$, ___ , ___

4) 0.07, 0.084, ___ , 0.112, ___

5) The rule is 'add $2\frac{4}{7}$'. The third term in the sequence is $4\frac{2}{7}$.
Find the first two terms and then find the fifth term.

6) The rule is 'subtract 1.085'. The third term in the sequence is 12.6.
Find the first two terms and then find the fifth term.

7) The tenth term in a sequence is 0.81. Can the first term be zero? Explain your thinking.

Solving problems about percentage, fraction and decimal equivalents

UNIT 20

National Curriculum link:

Solve problems which require knowing percentage and decimal equivalents of $\frac{1}{2}$, $\frac{1}{4}$, $\frac{1}{5}$, $\frac{2}{5}$, $\frac{4}{5}$ and those fractions with a denominator of a multiple of 10 or 25.

Year 5 pupils should already know that:

- Percentages, decimals and fractions are different ways of expressing proportions
- Per cent (%) relates to 'number of parts per hundred'
- A percentage can be represented as a fraction or a decimal, e.g. 25% = 0.25 = $\frac{25}{100}$ or $\frac{1}{4}$

Supporting understanding

Children should know that percentages, decimals and fractions are different ways of expressing proportions. It is important that the connections are always made so that they can confidently convert from one to the other.

Fraction bars and the relationship between a whole and 100% can be used to reinforce simple equivalents for halves, quarters and tenths. This can be further developed to consider eighths and fifths.

Whole = 100%			
$\frac{1}{2}$ = 50%		$\frac{1}{2}$ = 50%	
$\frac{1}{4}$ = 25%	$\frac{1}{4}$ = 25%	$\frac{1}{4}$ = 25%	$\frac{1}{4}$ = 25%

$\frac{1}{10}$=10%	$\frac{1}{10}$=10%	$\frac{1}{10}$=10%	$\frac{1}{10}$=10%	$\frac{1}{10}$=10%	$\frac{1}{10}$=10%	$\frac{1}{10}$=10%	$\frac{1}{10}$=10%	$\frac{1}{10}$=10%	$\frac{1}{10}$=10%

Percentage, decimal and fraction equivalents

Through previous work, children know that they can use multiplication facts to find equivalent fractions, e.g:

$$\frac{1}{5} = \frac{2}{10} = \frac{3}{15} = \frac{4}{20} = \frac{5}{25}$$

Knowing this will help us to find percentage equivalents for many fractions (when we are unsure of the decimal equivalent), as long as the denominator is a factor of 100. Of course, a calculator can also come in handy!

$\frac{1}{5} = \frac{20}{100}$ = 20% (or 0.2) as 5 multiplied by 20 is 100

$\frac{3}{25} = \frac{12}{100}$ = 12% (or 0.12) as 25 multiplied by 4 is 100

$\frac{7}{20} = \frac{35}{100}$ = 35% (or 0.35) as 20 multiplied by 5 is 100

ANSWERS

Task A: Children use strips of paper to find fraction and percentage equivalents

Task B: 1) Flour 30%, 0.3, $\frac{30}{100}$ ($\frac{3}{10}$); butter 15%, 0.15, $\frac{15}{100}$ ($\frac{3}{20}$); sugar 10%, 0.1, $\frac{10}{100}$ ($\frac{1}{10}$); apples 25%, 0.25, $\frac{25}{100}$ ($\frac{1}{4}$); raspberries 20%, 0.2, $\frac{20}{100}$ ($\frac{1}{5}$) **2)** Flour + raspberries; butter + sugar + apples **3)** Ami saw the digit 3 in 30% and thought this must be $\frac{1}{3}$; proof that 30% = $\frac{3}{10}$ and / or $\frac{1}{3}$ is 33.33%

Task C: 1) 27.5% **2)** 0.125 is equivalent to $\frac{1}{8}$, and he has confused $\frac{1}{8}$ with 8%; proof that e.g. 8% is $\frac{8}{100}$ so 0.08 not 0.125 **3)** 11% needed (16.5% + 11% = 27.5%), or $\frac{11}{100}$

In the classroom

Revisit the key point that a percentage can be represented as a fraction or a decimal.

Invite children to suggest fraction and percentage equivalents that they know.

Encourage children to use this language:

'I know that … is equal to …% because …'

Sometimes the fractions we want to represent as percentages are not written with a denominator of 100, e.g:

In a test, Pete scored $\frac{40}{50}$.

What do you notice? What do I need to do?

Establish that we can use equivalent fractions by looking at the denominator and its relationship to 100 (the denominator we need). In this case, we must multiply both parts of the fraction by 2 to give $\frac{80}{100}$ or 80%.

Ask children to practise this idea. Pose these, or similar, questions for different groups to consider:

- *In a test with 50 questions, Sami got 30 correct. What fraction will help to find his score as a percentage?*

- *In a test with 25 questions, Tom got 15 correct. What fraction will help to find his score as a percentage?*

- *How many questions will Ami need to answer correctly, in a test of 20 questions, to score the same as the others? How do you know?*

Now ask children to think about how we can use fractional equivalents of percentages to help to find decimal equivalents. Establish that a percentage such as 60% can be shown as $\frac{60}{100}$, and that by using the relationship between fractions and division, this can be calculated as 60 ÷ 100. Children should be reminded how they can use knowledge of place value and the effect of dividing by 100.

Task A (Independent task or guided learning with an adult)

Ask children to use strips of paper to support finding halves and quarters.

Whole = 100%			
$\frac{1}{2}$ = 50%		$\frac{1}{2}$ = 50%	
$\frac{1}{4}$ = 25%	$\frac{1}{4}$ = 25%	$\frac{1}{4}$ = 25%	$\frac{1}{4}$ = 25%

- Using knowledge of halving 100 and then halving again to find quarter, children create fraction and percentage equivalents.
- Encourage them to find percentage equivalents of 0.5, $\frac{2}{4}$ and $\frac{3}{4}$.
- Use a pre-prepared tenths strip and challenge children to find a way of calculating the percentage that each tenth represents.

Task B (Independent task or guided learning with an adult)

The list below shows the proportion of each ingredient needed to make a pie.

Raspberry and Apple Pie

Ingredients

Flour	30%
Butter	0.15
Sugar	0.1
Apples	25%
Raspberries	$\frac{20}{100}$

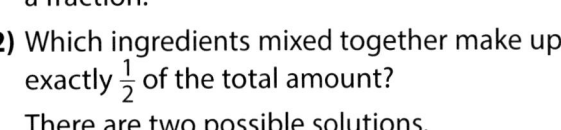

1) Write the amount of each of the ingredients as a percentage, a decimal and a fraction.

2) Which ingredients mixed together make up exactly $\frac{1}{2}$ of the total amount?

   There are two possible solutions.

3) Ami thinks that the flour makes up $\frac{1}{3}$ of the ingredients.

   Why does she think this?

   How can you prove that she has made a mistake?

Task C (Independent task)

The list below shows the proportion of each ingredient needed to make a pasta dish.

Pasta with vegetables

Ingredients

Pasta	$\frac{13}{25}$
Tomatoes	✳
Mushrooms	✳
Onions	0.125
Olives	8%

1) What percentage of the ingredients is made up of mushrooms and tomatoes?

2) Tom thinks that an equal amount of onions and olives is needed for this recipe.

   Why does he think this?

   How can you prove that he has made a mistake?

3) More tomatoes are needed in this recipe than mushrooms. The difference between the amounts is 5.5%.

   Find the amount of mushrooms needed and show this as a percentage and as a decimal.

UNIT 21 Solving problems using percentages, decimals and fractions

National Curriculum link:

[Non-statutory guidance] **Make connections between percentages, fractions and decimals (for example, 100% represents a whole quantity and 1% is $\frac{1}{100}$, 50% is $\frac{50}{100}$, 25% is $\frac{25}{100}$) and relate this to finding 'fractions of'.**

Year 5 pupils should already know that:

* Per cent (%) relates to 'number of parts per hundred'
* A percentage can be represented as a fraction or a decimal, e.g. 25% = 0.25 = $\frac{25}{100}$ or $\frac{1}{4}$
* We use division and multiplication to find a non-unit fraction of an amount or set of objects

Supporting understanding

Children will now be very familiar with the fraction bar image and should be able to use the structure of each bar to help to identify percentage equivalents.

100%									
50%					50%				
25%		25%		25%		25%			
10%	10%	10%	10%	10%	10%	10%	10%	10%	10%

It is important that 100% is shown as the whole (or 1) and that it is split into equal parts, e.g. into two equal parts of 50% so each is equal to $\frac{1}{2}$.

Calculating percentages of amounts

Children have been using resources and images to support finding fractions of amounts since Year 1, and for halving in the Early Years.

By developing these images further, we can use them to support finding simple percentages of amounts, e.g:

25% of £40 is £10 because $\frac{1}{4}\left(\frac{25}{100}\right)$ of £40 is £10

10% of £40 is £4 because $\frac{1}{10}\left(\frac{10}{100}\right)$ of £40 is £4

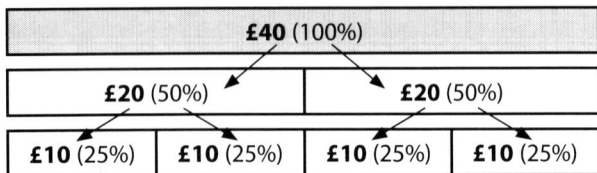

£40 (100%)									
£20 (50%)					£20 (50%)				
£10 (25%)		£10 (25%)		£10 (25%)		£10 (25%)			

£40 (100%)									
£4 (10%)	£4 (10%)	£4 (10%)	£4 (10%)	£4 (10%)	£4 (10%)	£4 (10%)	£4 (10%)	£4 (10%)	£4 (10%)

In the classroom

Remind children of the fraction bar images they have used previously (left).

Ask children to use the images to come up with a range of fraction and percentage equivalents by explaining how much of each bar is represented, e.g:

I know that 75% is equivalent to $\frac{3}{4}$ because each quarter is 25% or $\frac{25}{100}$.

Revisit finding unit and non-unit fractions of amounts.

Ask children to discuss strategies they would use to find e.g:

$\frac{1}{3}$ of 150 g, $\frac{7}{10}$ of 30 m, $\frac{3}{5}$ of £60 or $\frac{5}{12}$ of an hour.

If not already suggested, return to the fraction bar model children have used in previous units.

How can we use the fraction bars to help us to find: 50% and 25%, 75% and 10%, 20% and 30% of £40?

Writing £40 as the whole, ask different groups to consider each pair of percentages.

Use the model on the left to confirm percentage amounts. Use this language: '10% of £40 is £4 because $\frac{1}{10}\left(\frac{10}{100}\right)$ of £40 is £4.'

Tell children about some other percentages you have been finding, this time of £20:

The amounts I found are £5, £15 and £6. What percentages of £20 are these?

Ask more able children to focus on £6.

ANSWERS

Task A: 1) £8 **2)** £4 **3)** £12 **4)** £5 **5)** £2.50 **6)** £7.50 **7)** £12 because 75% is the same as finding $\frac{3}{4}$ **8)** £2.50 because 25% is the same as finding $\frac{1}{4}$ **Challenge:** £15

Task B: 1) £60 **2)** £30 **3)** £90 **4)** £12 **5)** £21 **6)** £10.50 **7)** £31.50 **8)** £4.20 **9)** £4.20 because 10% is the same as finding $\frac{1}{10}$ **Challenge:** 20% is double 10% and double £12 is £24

Task C: 1) £22.50 **2)** £67.50 **3)** £9 **4)** £27 **5)** £7.80 **6)** £11.70 **7)** £1.56 **8)** £1.56 **9)** 5% is half of 10%, but as we have double the amount, the answers / proportions are the same **Challenge:** E.g. 50% of £45 (1), 37.5% of £180 (2), 5% of £180 or 20% of £45 (3), 15% of £180 (4), 25% of £31.20 (5), 25% of £46.80 (6)

Task A (Independent task)

Use the fraction bars to help you to find the following percentages of each amount.

100%			
50%		50%	
25%	25%	25%	25%

1) 50% of £16
2) 25% of £16
3) 75% of £16
4) 50% of £10
5) 25% of £10
6) 75% of £10

7) Which of your answers is the same as finding $\frac{3}{4}$ of £16? Why?

8) Which of your answers is the same as finding $\frac{1}{4}$ of £10? Why?

Challenge:

Can you find 25% of £60?

HINT: Remember that finding $\frac{1}{4}$ is the same as finding a half of $\frac{1}{2}$.

Task B (Independent task)

Use the fraction bars to help you to find the following percentages of each amount.

100%									
50%					50%				
25%		25%		25%		25%			
10%	10%	10%	10%	10%	10%	10%	10%	10%	10%

1) 50% of £120
2) 25% of £120
3) 75% of £120
4) 10% of £120
5) 50% of £42
6) 25% of £42
7) 75% of £42
8) 10% of £42

9) Which of your answers is the same as finding $\frac{1}{10}$ of £42? Why?

Challenge:

How would you find 20% of £120?

Task C (Independent task)

100%									
50%					50%				
25%		25%		25%		25%			
10%	10%	10%	10%	10%	10%	10%	10%	10%	10%

Find the following percentages of each amount:

1) 25% of £90
2) 75% of £90
3) 10% of £90
4) 30% of £90
5) 50% of £15.60
6) 75% of £15.60
7) 10% of £15.60
8) 5% of £31.20

9) What do you notice about the answers to questions 7 and 8? Why did this happen?

Challenge:

Use this idea to help you to find other percentages of different amounts that will give the same answers as questions 1 to 6.

UNIT 22 Solving problems about fractions, decimals and percentages

National Curriculum link:

Solve problems which require knowing percentage and decimal equivalents of $\frac{1}{2}, \frac{1}{4}, \frac{1}{5}, \frac{2}{5}, \frac{4}{5}$ and those fractions with a denominator of a multiple of 10 or 25.

Year 5 pupils should already know that:

- Percentages, decimals and fractions are different ways of expressing proportions
- Per cent (%) relates to 'number of parts per hundred'
- A percentage can be represented as a fraction or a decimal, e.g. 25% = 0.25 = $\frac{25}{100}$ or $\frac{1}{4}$

Supporting understanding

Reasoning about number lines or scales presented in different orientations requires children to think about what they know and how they can apply it to the situation.

We can use measurement or statistics as contexts for a range of different problems.

Language structures will continue to support reasoning and help to deepen understanding.

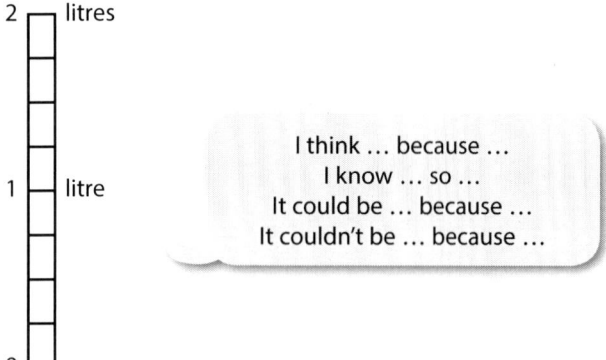

I think … because …
I know … so …
It could be … because …
It couldn't be … because …

Making lists to organise results

In the problem to be developed in the classroom, we initially know that Symi has 25% of a litre more than Tim. We also know that Tim has more than $\frac{2}{5}$ litre and that the amount can be written to one decimal place.

A table will help to organise possibilities:

Tim	Symi
500 ml	750 ml
600 ml	850 ml
700 ml	950 ml
etc.	etc.

As the problem develops, solutions that are no longer possible can be discarded.

In the classroom

Using the scale (left), ask children to discuss and explain where these proportions of a litre would be placed and to give their value in millilitres:

25%   1.4 $\frac{3}{4}$ 10% $\frac{3}{10}$ 1.125

Agree positions and values for some of the proportions.

Pose the following problem:

Tim and Symi each have some orange squash in their 2 litre measuring jugs.

Symi has 25% of a litre more than Tim.

Tim's amount can be written to one decimal place, and he has more than $\frac{2}{5}$ litre.

Ask children to work in groups to suggest possible values in ml, e.g. 700 ml for Tim so Symi has 950 ml.

Work with those who need more support, giving suggested amounts for Tim, as necessary.

Pose these, or similar, questions for groups to consider:

- *What do we know about the number of ml that Tim has? (Multiple of 100 ml.) Why?*
- *What is the least amount that Tim can have? Why?*

Discuss ways to record results and consider the use of a table to organise solutions (as shown to the left).

Symi has more than 0.925 litres of orange squash.

Tim has less than $1\frac{1}{5}$ litres of orange squash.

Ask children to consider the solutions that are still possible and those that are not possible and to explain their thinking.

The problem develops in the independent tasks, but note that the build-up to this point of the lesson, with children working in groups, should be flexible and will take longer than usual.

ANSWERS

Task A: 1) $\frac{9}{10}$ litre **2)** $2\frac{1}{4}$ litres **3)** $1\frac{350}{1000}$ litres **4)** Solution 1

Task B: 1) Solution 1 **2)** $\frac{700}{1000}$ litre and $\frac{950}{1000}$ litre **3)** 0.6 litre

Task C: 1) Solution 1 **2)** Tim 70% and Symi 95% **3)** 0.325 litre

Task A (Independent task)

We still do not know exactly how many litres of orange squash Symi and Tim have.

There are five possible solutions now:

Solution	Tim	Symi
1	700 ml	0.95 litres
2	800 ml	1.05 litres
3	900 ml	1.15 litres
4	1 litre	1.25 litres
5	1.1 litres	1.35 litres

1) How many tenths of a litre does Tim have in Solution 3?

2) Add together 1 litre and 1.25 litres from Solution 4. Write the amount as a mixed number.

3) In Solution 5, Tim can write his amount as $1\frac{100}{1000}$ litres. How can Symi write hers in a similar way?

4) Symi and Tim pour all their squash into one measuring jug. They have 1650 ml altogether. Which solution is this?

Task B (Guided learning with an adult)

We still do not know exactly how many litres of orange squash Symi and Tim each have in their 2 litre jugs.

There are five possible solutions now:

Solution	Tim	Symi
1	700 ml	0.95 litres
2	800 ml	1.05 litres
3	900 ml	1.15 litres
4	1 litre	1.25 litres
5	1.1 litres	1.35 litres

1) Use the clues to find the correct solution.

CLUES:
- Tim's amount cannot be written as a multiple of 0.2 litres.
- Symi's amount cannot be written as the mixed number $1\frac{1}{4}$ litres.
- Tim has less than 100% of a litre.
- Together, they have less than $\frac{18}{10}$ litre.

2) Write the amounts as fractions with denominator 1000.

3) Their friend Abi has 10% of a litre less than Tim. Write the amount of squash Abi has as a decimal.

Task C (Independent task)

We still do not know exactly how many litres of orange squash Symi and Tim each have in their 2 litre jugs.

There are five possible solutions now:

Solution	Tim	Symi
1	700 ml	0.95 litres
2	800 ml	1.05 litres
3	900 ml	1.15 litres
4	1 litre	1.25 litres
5	1.1 litres	1.35 litres

1) Use the clues to find the correct solution.

CLUES:
- Tim's amount cannot be written as a fraction in fifths of a litre.
- Symi's amount can't be written as the improper fraction $\frac{1250}{1000}$ litres.
- Tim's amount can't be written as any improper fraction of a litre.
- Symi's squash is 5% of a litre more or 5% of a litre less than half the jug.

2) Write the amounts as percentages of a litre.

3) Their friend Paulo has $\frac{3}{8}$ litre of squash less than Tim. Write the amount of squash Paulo has as a decimal.

RISING ☆ STARS
Maths

Supporting schools through curriculum change

www.risingstars-uk.com/maths